GOD

GOD

THE FAILED HYPOTHESIS

How Science Shows
That God
Does Not Exist

VICTOR J. STENGER

Prometheus Books

59 John Glenn Drive
Amherst, New York 14228-2197

Published 2007 by Prometheus Books

Inquiries should be addressed to
Prometheus Books
59 John Glenn Drive
Amherst, New York 14228-2197
VOICE: 716-691-0133, ext. 207
FAX: 716-564-2711
WWW.PROMETHEUSBOOKS.COM

11 10 09 08 07   5 4 3 2

Library of Congress Cataloging-in-Publication Data

Stenger, Victor J., 1935–
    God : the failed hypothesis : how science shows that God does not exist /
Victor J. Stenger.
        p. cm.
    Includes bibliographical references and index.
    ISBN 978-1-59102-481-1 (alk. paper)
    1. Religion and science. 2. God—Proof. 3. Atheism. I. Title.

BL240.3.S738 2007
212'.1—dc22

                                                                    2006032867

Printed in the United States of America on acid-free paper

TABLE OF CONTENTS

ACKNOWLEDGMENTS

A s with my previous books, I have benefited greatly from comments and corrections provided by a large number of individuals with whom I communicate regularly on the Internet. I am particularly grateful for the substantial suggestions provided by Bill Benson, Celel Berker, Eleanor Binnings, Richard Carrier, Lawrence Crowell, Yonathan Fishman, Joseph Hilbe, James Humpreys, Ludwig Krippahl, Andrew Laska, Justin Lloyd, Ken MacVey, Don McGee, Brent Meeker, Anne O'Reilly, Loren Petrich, Kerry Regier, John Syriatowicz, Phil Thrift, George Tucker, Ed Weinmann, Roahn Wynar, and Bob Zanelli.

I must also acknowledge the support and valuable comments provided to me by Jeff Dean of Blackwell Publishing and his three anonymous reviewers. I have implemented many of their suggestions in this manuscript.

And, as with my previous books, I have enjoyed the support

of Paul Kurtz and Steven L. Mitchell and their dedicated staff at Prometheus Books.

Of course, none of my work would be possible without the help and encouragement of my wife Phylliss and our two children.

PREFACE

The sciences do not try to explain, they hardly even try to interpret, they mainly make models. By a model is meant a mathematical construct which, with the addition of certain verbal interpretations, describes observed phenomena. The justification of such a mathematical construct is solely and precisely that it is expected to work.

—John von Neumann[1]

VIEW FROM THE SIDELINES

Throughout history, arguments for and against the existence of God have been largely confined to philosophy and theology. In the meantime, science has sat on the sidelines and quietly watched this game of words march up and down the field. Despite the fact that science has revolutionized every aspect of human life and greatly clarified our understanding of the world, somehow the notion has arisen that it has nothing to say about a supreme being that much of humanity worships as the source of all reality.

In his 1999 book, *Rocks of Ages*, famed paleontologist Stephen Jay Gould referred to science and religion as two "non-overlapping magisteria," with science concerning itself with understanding the natural world while religion deals with issues of morality.[2] However, as many reviewers pointed out, this amounted to a redefinition of religion as moral philosophy. In fact, most religions do more than simple moralizing but make basic pronouncements about nature, which science is free to evaluate. Furthermore, science has an obvious role in the study of physical objects, such as the Shroud of Turin, which may have religious implications. And, why can't science consider moral issues, which involve observable and sometimes even quantifiable human behavior?

In a poll taken in 1998, only 7 percent of the members of the US National Academy of Sciences, the elite of American scientists, said they believed in a personal God.[3] Nevertheless, most scientists seem to prefer as a practical matter that science should stay clear of religious issues. Perhaps this is a good strategy for those who wish to avoid conflicts between science and religion, which might lead to less public acceptance of science, not to mention that most dreaded of all consequences—lower funding. However, religions make factual claims that have no special immunity from being examined under the cold light of reason and objective observation.

Besides, scientific arguments *for* the existence of God, that is, arguments based on observations rather than authority, have been made since ancient times—as early as 77 BCE by Marcus Tullius Cicero (d. 43 BCE) in his work *De Natura Deorum* (*On the Nature of the Gods*).[4] Particularly influential was William Paley (d. 1805) with his *Natural Theology or Evidences of the Existence and Attributes of the Deity Collected from the Appearance of Nature*, first published in 1802.[5] In more recent years, theologians and theistic scientists have begun looking to science to provide support for their beliefs in a supreme being. Many books have been published purporting that modern theoretical and empirical science supports the proposition that God exists, and the popular media have been quick to promulgate this view.[6] Very few books or media stories have directly challenged that assertion. But if scientific arguments for the existence of God are to be allowed into intellectual discourse, then those against his existence also have a legitimate place.

In my 2003 book, *Has Science Found God?* I critically examined the claims of scientific evidence for God and found them inadequate.[7] In the present book, I will go much further and argue that by this moment in time science has advanced sufficiently to be able to make a definitive statement on the existence or nonexistence of a God having the attributes that are traditionally associated with the Judeo-Christian-Islamic God.

We now have considerable empirical data and highly successful scientific models that bear on the question of God's existence. The time has come to examine what those data and models tell us about the validity of the God hypothesis.

To be sure, the Judeo-Christian-Islamic God is not well defined. Not only do different views of God exist among these faiths, but also many differences can be found within each faith itself—between theologians and lay believers as well as from sect to sect. I will focus on those attributes of the God that the bulk of believers in each of these varied groups worship. Some of these attributes are also shared by the deities of religions outside the three great monotheisms.

I am well aware that sophisticated theologians have developed highly abstracted concepts of a god that they claim is consistent with the teachings of their faiths. One can always abstract any concept so it is out of the realm of scientific investigation. But these gods would not be recognized by the typical believer.

In the three monotheisms, God is viewed as a supreme, transcendent being—beyond matter, space, and time—and yet the foundation of all that meets our senses that is described in terms of matter, space, and time. Furthermore, this God is not the god of deism, who created the world and then left it alone, or the god of pantheism, who is equated with all of existence. The Judeo-Christian-Islamic God is a nanosecond-by-nanosecond participant in each event that takes place in every cubic nanometer of the universe, from the interactions of quarks inside atomic nuclei to the evolution of stars in the most distant galaxies. What is more, God listens to every thought and participates in each action of his very special creation, a minute bit of organized matter called humanity that moves around on the surface of a tiny pebble in a vast universe.

So, when I use uppercase G, I mean the Judeo-Christian-Islamic God. Other gods will be lowercase. I will also use the traditional masculine pronouns in referring to God. This book is an investigation of the evidence for the existence of God—not all gods. It might be likened to a physicist investigating the existence of a massless charged particle, but not all particles.

SUPERNATURAL SCIENCE

No consensus exists among philosophers of science on what distinguishes science from pseudoscience or nonscience, although most scientists would say they know pseudoscience when they see it. In this book, I will take science to refer to the performing of objective observations by eye and by instrument and the building of models to describe those observations. These models are not simple snapshots of the observations, but they utilize elements and processes or mechanisms that attempt to be universal and general so that not only one set of observations is described but also all the observations that fit into as wide a class as possible. They need not always be mathematical, as asserted by John von Neumann in the epigraph to this chapter.

Perhaps the most outstanding current (mathematical) example is the *standard model of elementary particles and forces* in which all of familiar matter is composed of just three particles: the *up quark*, the *down quark*, and the electron. This model was formulated in the 1970s and to this date remains consistent with all the measured properties of matter made in our most sophisticated laboratories on Earth and observed in space with our most powerful telescopes.

Notice that the main purpose of scientific models is to *describe* rather than *explain*. That is, they are deemed successful when they agree with all observations, especially those that would have falsified the model had those observations turned out otherwise. Often this process takes the form of *hypothesis testing*, in which a model is proposed as a series of hypotheses that are then tested against carefully controlled observations. Whether the elements and processes that make up a successful model are to be taken as intrinsic parts of reality is not a question that can be simply answered since we can never know that the model might be falsified in the future. However, when a model is falsified, we can reasonably assume that those elements and processes that are unique to the model and not also part of another, successful model are likely not intrinsic parts of reality.

My analysis will be based on the contention that God should be detectable by scientific means simply by virtue of the fact that he is supposed to play such a central role in the operation of the universe and the lives of humans. Existing scientific models contain no place where God is included as an ingredient in order to describe observations. Thus, if God exists, he must appear somewhere within the gaps or errors of scientific models.

Indeed, the "God of the gaps" has long been a common argument for God. Science does not explain everything, so there is always room for other explanations and the believer is easily convinced that the explanation is God. However, the God of the gaps argument by itself fails, at least as a scientific argument, unless the phenomenon in question is not only currently scientifically inexplicable but can be shown to forever defy natural description. God can only show up by proving to be necessary, with science equally proven to be incapable of providing a plausible account of the phenomenon based on natural or material processes alone.

This may strike the reader as an impossible requirement. How can we ever know that science will never be able to provide a "natural" account for some currently mysterious phenomenon? I claim this is within the realm of possibility, if not with 100 percent certainty, within a reasonable doubt. Using the historical association of *natural* with *material,* I will provide hypothetical examples of phenomena that, if observed, cannot be of material origin beyond a reasonable doubt. Since by all accounts God is nonmaterial, his presence would be signaled, beyond a reasonable doubt, by the empirical verification of such phenomena.

Some scientists have raised objections to the association of natural with material. They say all observable phenomena are "natural," by (their) definition. Others say any testable theory is "natural," by (their) definition. I prefer not to indulge in endless arguments over the meanings of words that never seem to converge on a consensus. I have stated how I will use the words *natural* and *supernatural,* as synonymous with *material* and *nonmaterial.* The supernatural cannot be banished from science by mere definition.

I define matter as anything that kicks back when you kick it. It is the stuff of physics. By "kick" I refer to the universal observation process in which particles, such as the photons that compose light, are bounced off objects. Measurements on the particles that bounce back into our eyes and other sensors give us properties of the observed object called mass, momentum, and energy that we identify with matter. Those measurements are described with models that contain purely material processes—the dynamical principles of physics—all subject to empirical testing and falsification.[8]

Many scientists will object that the supernatural or nonmaterial cannot be tested in any analogous manner. Indeed, in recent political battles in the United States that have pit science against conservative religious groups who see their beliefs threatened by evolution, prominent scientists and national science organizations have made public statements and given court testimony to the effect that science can only deal with natural causes. In this they have played right into the hands of those who try to argue that science has a dogmatic commitment to materialism that prevents it from even considering any alternatives.

In this book I will show that a number of proposed supernatural or nonmaterial processes are empirically testable using standard scientific methods. Furthermore, such research is being carried out by reputable scientists associated with reputable institutions and published in reputable scientific journals. So the public statements by some scientists and their national organizations that science has nothing to do with the supernatural are belied by the facts.

True that science generally makes the assumption called *methodological naturalism*, which refers to the self-imposed convention that limits inquiry to objective observations of the world and generally (but, as we will see, not necessarily) seeks natural accounts of all phenomena. This is often confused with *metaphysical naturalism*, which assumes that reality itself is purely natural, that is, composed solely of material objects. While it cannot be denied that most physical scientists, at least, think this is the case, they cannot prove it. Furthermore, they have no need to try since ultimately it is not a scientific question amenable to empirical adjudication. If it were, it would be physics and not metaphysics.

In this book I will show that certain natural, material phenomena are implied by the God hypothesis. The observation of any of these phenomena would defy all reasonable natural, material descriptions.

Despite philosophical and historical literature in the past century that described the history of science as a series of revolutions and "paradigm shifts,"[9] the fundamental notion of matter and material processes has not been changed since the time of

Newton—only embellished.[10] Anything that can be shown to violate those principles, to have properties different from those long associated with matter, would be of such world-shaking significance that, for want of a better term, we could call them supernatural.

As far as we can tell from current scientific knowledge, the universe we observe with our senses and scientific instruments can be described in terms of matter and material processes alone. Certainly scientists will initially search for a material account of any new phenomenon since parsimony of thought requires that we seek the simplest models first, those that make the fewest new, untried hypotheses. However, should all material explanations fail, there is nothing stopping the empirical testing of hypotheses that go beyond those of conventional physical science.

GAPS FOR GOD?

Well aware that the existence of God is not proved from the incompleteness of science alone, some theologians and theistic scientists are now claiming that they have uncovered gaps in scientific theories that can only be filled by a supreme being operating outside the natural realm. They boldly assert that science cannot account for certain phenomena and, furthermore, never will. The new "proofs" are based on claims that the complexity of life cannot be reduced, and never will be reduced, to purely natural (material) processes. They also assert that the constants and laws of physics are so fine-tuned that they cannot have come about naturally, and that the origin of the physical universe and the laws it obeys cannot have "come from nothing" without supernatural intervention. Believers also cite results from purported carefully controlled experiments that they say provide empirical evidence for a world beyond matter that cannot be accounted for by material processes alone.

In order to estimate effectively the credibility of these claims, we must be careful to properly locate the burden of proof. That burden rests on the shoulders of those who assert that science will never be able to account naturally for some phenomenon, that is, describe the phenomenon with a model containing only material elements and processes. If a plausible scientific model consistent with all existing knowledge can be found, then the claim fails. That model need not be proven to be correct, just not proven to be incorrect.

If we can find plausible ways in which all the existing gaps in scientific knowledge one day may be filled, then the scientific arguments for the existence of God fail. We could then conclude that God need not be included in the models we build to describe phenomena currently observable to humans. Of course, this leaves open the possibility that a god exists that is needed to account for phenomena outside the realm of current human observation. He might show up in some future space expedition, or in some experiment at a giant particle accelerator. However, that god would not be a god who plays an important role in human life. It is not God.

EXAMINING THE EVIDENCE AGAINST GOD

Evaluating the arguments that science has uncovered evidence for God is only part of my task, which was largely completed in *Has Science Found God?* My primary concern here will be to evaluate the less familiar arguments in which science provides evidence *against* the existence of God.

The process I will follow is the scientific method of hypothesis testing. The existence of a God will be taken as a scientific hypothesis and the consequences of that hypothesis searched for in objective observations of the world around us. Various models will be assumed in which God has specific attributes that can be tested empirically. That is, if a God with such attributes exists, certain phenomena should be observable. Any failure to pass a specific test will be regarded as a failure of that particular model. Furthermore, if the actual observations are as expected in the absence of the specified deity, then this can be taken as an additional mark against his existence.

Where a failure occurs, the argument may be made that a hidden God still may exist. While this is a logically correct statement, history and common experience provide many examples where, ultimately, absence of evidence became evidence of absence. Generally speaking, when we have no evidence or other reason for believing in some entity, then we can be pretty sure that entity does not exist.[11] We have no evidence for Bigfoot, the Abominable Snowman, and the Loch Ness Monster, so we do not believe they exist. If we have no evidence or other reason for believing in God, then we can be pretty sure that God does not exist.

NOTES

1. As quoted in J. Tinsley Oden, acceptance remarks, 1993 John von Neumann Award Winner, *United States Association of Computational Mechanics Bulletin* 6, no. 3 (September 1993). Online at http://www.usacm.org/Oden's_acceptance_remarks.htm (accessed February 22, 2005).

2. Stephen J. Gould, *Rock of Ages: Science and Religion in the Fullness of Life* (New York: Ballantine, 1999).

3. Edward J. Larson and Larry Witham, "Leading Scientists Still Reject God," *Nature* 394 (1998): 313.

4. Marcus Tullius Cicero, *De Natura Deorum* or *On the Nature of the Gods*, ed. and trans. H. Rackham (New York: Loeb Classical Library, 1933).

5. William Paley, *Natural Theology or Evidences of the Existence and Attributes of the Deity Collected from the Appearance of Nature* (London: Halliwell, 1802).

6. Sharon Begley, "Science Finds God," *Newsweek*, July 20, 1998.

7. Victor J. Stenger, *Has Science Found God? The Latest Results in the*

Search for Purpose in the Universe (Amherst, NY: Prometheus Books, 2003). See references therein for the original claims.

8. Victor J. Stenger, *The Comprehensible Cosmos: Where Do the Laws of Physics Come From?* (Amherst, NY: Prometheus Books, 2006). Contains a complete discussion of the nature of matter and other physical entities.

9. Thomas Kuhn, *The Structure of Scientific Revolutions* (Chicago: University of Chicago Press, 1970).

10. Steven Weinberg, "The Revolution That Didn't Happen," *New York Review of Books*, October 8, 1998.

11. Keith Parsons, *God and the Burden of Proof: Platinga, Swinburne, and the Analytical Defense of Theism* (Amherst, NY: Prometheus Books, 1989).

Chapter 1

MODELS AND METHODS

All that belongs to human understanding, in this deep ignorance and obscurity, is to be skeptical, or at least cautious; and not to admit of any hypothesis, whatsoever; much less, of any which is supported by no appearance of probability.

—David Hume

LACK OF EVIDENCE

Many theologians and theistic scientists claim that evidence has been found for the existence of the Judeo-Christian-Islamic God or, at least, some being with supernatural powers. However, they cannot deny that their evidence is not sufficiently convincing to satisfy the majority of scientists. Indeed, as we saw in the preface, the overwhelming majority of prominent

American scientists has concluded that God does not exist. If God exists, where is he? Philosopher Theodore Drange has termed this the *lack-of-evidence argument*, which he states formally as follows:

1. Probably, if God were to exist, then there would be good objective evidence for his existence.
2. But there is no good objective evidence for his existence.
3. Therefore, probably God does not exist.

Drange criticizes premise 1 of the lack-of-evidence argument, pointing out that God could simply choose not to use the channel of objective evidence but directly implant that knowledge in human minds.[1] However, as he and others have pointed out, such a deity would not be a perfectly loving God and the very existence of nonbelievers in the world who have not resisted such belief is evidence against his existence.[2] The *problem of divine hiddenness* is one that has taxed the abilities of theologians over the years—almost as much as the *problem of evil*, which questions how an omnibenevolent, omnipotent, and omniscient God can allow so much unnecessary suffering among the planet's humans and animals. We will return to each of these problems.

However, independent of the unknowable intentions of a hypothetical being of infinite power and wisdom, objective evidence for an entity with godlike attributes should be readily available. After all, God is supposed to play a decisive role in every happening in the world. Surely we should see some sign of that in objective observations made by our eyes and ears, and especially by our most sensitive scientific instruments.

The founders and leaders of major religions have always claimed that God can be seen in the world around us. In Romans 1:20, St. Paul says: "Ever since the creation of the world his invisible nature, namely his eternal power and deity, have been clearly perceived in the things that have been made." We will look for evidence of God in the things that have been made.

THE NATURE OF SCIENTIFIC EVIDENCE

Before examining specific data, let us consider what constitutes "scientific evidence." Here I will limit myself to the kind of evidence that is needed to establish the validity of an extraordinary claim that goes beyond existing knowledge. Clearly the standard for this must be set much higher than that for an ordinary claim.

For example, an ordinary claim might be that an 81-milligram aspirin taken daily will reduce the chance of heart attacks and strokes. Such a claim is ordinary, because we have a plausible mechanism for such an effect in the resulting slight thinning of the blood. By contrast, an extraordinary claim might be that such a therapy would cure AIDS. Lacking any plausible mechanism, we would have to demand far more confirmatory data than in the first case.

We often hear of stories citing examples of dreams that came true. This would seem to suggest a power of the mind that goes beyond known physical capabilities. However, in this case, a strong selection process is taking place whereby all the millions of dreams that do not come true are simply ignored. Unless otherwise demonstrated, a plausible explanation that must first be ruled out is that the reported dream came true by chance selection out of many that had no such dramatic outcome.

How can we rule out chance or other artifacts? This is what the scientific method is all about. We might do a controlled experiment with hundreds of subjects recording their dreams upon awaking every morning. Independent investigators, with no stake in the outcome one way or another, would then perform a careful statistical analysis of the data. It would help if the dream outcomes were something simple and quantitative, like the winning number for a future lottery. Then the results could be compared with the easily calculated expectations from chance.

Allow me to list a few of the rules that the scientific community conventionally applies when evaluating any extraordinary claim. This is not complete by any means; nowhere can we find a document that officially lays down the scientific method to the complete satisfaction of a consensus of scientists and philosophers. However, five conditions suffice for our evaluation of claims of empirical evidence for extraordinary empirical claims in science:

Conditions for Considering Extraordinary Claims

1. The protocols of the study must be clear and impeccable so that all possibilities of error can be evaluated. The investigators, not the reviewers, carry the burden of identifying each possible source of error, explaining how it was minimized, and providing a quantitative estimate of the effect of each error. These errors can be systematic—attributable to biases in the experimental set up—or statistical—the result of chance fluctuations. No new effect can be claimed unless all the errors are small enough to make it highly unlikely that they are the source of the claimed effect.

2. The hypotheses being tested must be established clearly and explicitly before data taking begins, and not changed midway through the process or after looking at the data. In particular, "data mining" in which hypotheses are later changed to agree with some interesting but unanticipated results showing up in the data is unacceptable. This may be likened to painting a bull's-eye around wherever an arrow has struck. That is not to say that certain kinds of exploratory observations, in astronomy, for example, may not be examined for anomalous phenomena. But they are not used in hypothesis testing. They may lead to new hypotheses, but these hypotheses must then be independently tested according to the protocols I have outlined.

3. The people performing the study, that is, those taking and analyzing the data, must do so without any prejudgment of how the results should come out. This is perhaps the

most difficult condition to follow to the letter, since most investigators start out with the hope of making a remarkable discovery that will bring them fame and fortune. They are often naturally reluctant to accept the negative results that more typically characterize much of research. Investigators may then revert to data mining, continuing to look until they convince themselves they have found what they were looking for.[3] To enforce this condition and avoid such biases, certain techniques such as "blinding" may be included in the protocol, where neither the investigators nor the data takers and analyzers know what sample of data they are dealing with. For example, in doing a study on the efficacy of prayer, the investigators should not know who is being prayed for or who is doing the praying until all the data are in and ready to be analyzed.

4. The hypothesis being tested must be one that contains the seeds of its own destruction. Those making the hypothesis have the burden of providing examples of possible experimental results that would falsify the hypothesis. They must demonstrate that such a falsification has not occurred. A hypothesis that cannot be falsified is a hypothesis that has no value.

5. Even after passing the above criteria, reported results must be of such a nature that they can be independently replicated. Not until they are repeated under similar conditions by different (preferably skeptical) investigators will they be finally accepted into the ranks of scientific knowledge.

Our procedure in the following chapters will be to select out, one by one, certain limited sets of attributes and examine the empirical consequences that can reasonably be expected by the hypothesis of a god having those attributes. We will then look for evidence of these empirical consequences.

FALSIFICATION

Falsification was the demarcation criterion proposed in the 1930s by philosophers Karl Popper[4] and Rudolf Carnap[5] as a means for distinguishing legitimate scientific models from nonscientific conjectures. Since then, however, philosophers of science have found falsification insufficient for this purpose.[6] For example, astrology is falsifiable (indeed, falsified) and not accepted as science. Nevertheless, falsification remains a very powerful tool that is used whenever possible. When a hypothesis is falsifiable by a direct empirical test, and that test fails, then the hypothesis can be safely discarded.

Now, a certain asymmetry exists when testing scientific models. While failure to pass a required test is sufficient to falsify a model, the passing of the test is not sufficient to verify the model. This is because we have no way of knowing a priori that other, competing models might be found someday that lead to the same empirical consequences as the one tested.

Often in science, models that fail some empirical test are modified in ways that enable them to pass the test on a second or third try. While some philosophers have claimed this shows that falsification does not happen in practice, the modified model can be regarded as a new model and the old version was still falsified. I saw many proposed models falsified during my forty-year research career in elementary particle physics and astrophysics; it does happen in practice.[7]

Popper restricted falsification (which he equates to *refutability*) to empirical statements, and declared, "philosophical theories, or metaphysical theories, will be *irrefutable by definition*."[8] He also noted that certain empirical statements are irrefutable. These are statements that he called "strict or pure existential statements." On the other hand, "restricted" existential statements are refutable. He gives this example:

"There exists a pearl which is ten times larger than the next largest pearl." If in this statement we restrict the words "There exists" to some finite region in space and time, then it may of course become a refutable statement. For example, the following statement is obviously empirically refutable: "At this moment and in this box here there exist at least two pearls one of which is ten times larger than the next largest pearl in this box." But then this statement is no longer a strict or pure existential statement: rather it is a *restricted* existential statement. A strict or pure existential statement applies to the whole universe, and it is irrefutable simply because there can be no method by which it could be refuted. For even if we were able to search our entire universe, the strict or pure existential statement would not be refuted by our failure to discover the required pearl, seeing that it might always be hiding in a place where we are not looking.[9]

By this criterion, it would seem that the existence of God cannot be empirically refuted because to do so would require making an existential statement applying to the whole universe (plus whatever lies beyond). But, in looking at Popper's example, we see this is not the case for God. True, we cannot refute the existence of a God who, like the pearl in Popper's example, is somewhere outside the box, say, in another galaxy. But God is supposed to be everywhere, including inside every box. So when we search for God inside a single box, no matter how small, we should either find him, thus confirming his existence, or not find him, thus refuting his existence.

CAN SCIENCE STUDY THE SUPERNATURAL?

Most national science societies and organizations promoting science have issued statements asserting that science is limited to the consideration of natural processes and phenomena. For example, the United States National Academy of Sciences has stated, "Science is a way of knowing about the natural world. It is limited to explaining the natural world through natural causes. Science can say nothing about the supernatural. Whether God exists or not is a question about which science is neutral."[10]

Those scientists and science organizations that would limit science to the investigation of natural causes provide unwitting support for the assertion that science is dogmatically naturalistic. In a series of books in the 1990s, law professor Phillip Johnson argued that the doctrine that nature is "all there is" is the virtually unquestioned assumption that underlies not only natural science but intellectual work of all kinds.[11] In many of the public discussions we hear today, science is accused of dogmatically refusing to consider the possible role other than natural processes may play in the universe.

Given the public position of many scientists and their organizations, Johnson and his supporters have some basis for making a case that science is dogmatically materialistic. However, any type of dogmatism is the very antithesis of science. The history of science, from Copernicus and Galileo to the present, is replete with examples that belie the charge of dogmatism in science.

What history shows is that science is very demanding and does not blindly accept any new idea that someone can come up with. New claims must be thoroughly supported by the data, especially when they may conflict with well-established knowledge. Any research scientist will tell you how very difficult it is to discover new knowledge, convince your colleagues that it is correct—as they enthusiastically play devil's advocate—and then get your results through the peer-review process to publication. When scientists express their objections to claims such as evidence for intelligent design in the universe, they are not being dogmatic. They are simply applying the same standard they would for any other extraordinary claim and demanding extraordinary evidence.

Besides, why would any scientist object to the notion of intelligent design or other supernatural phenomena, should the data warrant that they deserve attention? Most scientists would be delighted at the opening up of an exciting new field of study that would undoubtedly receive generous funding. As we will see, intelligent design, in its current form, simply incorporates neither the evidence nor the theoretical arguments to warrant such attention.

Furthermore, the assertions that science does not study the supernatural and that supernatural hypotheses are untestable are factually incorrect. Right under the noses of the leaders of national science organizations who make these public statements, capable, credentialed scientists are investigating the possibility of supernatural causes. As we will discuss in a later chapter, reputable institutions such as the Mayo Clinic, Harvard University, and Duke University are studying phenomena that, if verified, would provide strong empirical support for the existence of some nonmaterial element in the universe. These experiments are designed to test the healing power of distant, blinded intercessory prayer. Their results have been published in peer-reviewed medical journals.

Unfortunately, the prayer literature is marred by some very poor experimental work. But in reading the best of the published papers of the most reputable organizations you will witness all the indications of proper scientific methodology at work. If they are not science, then I do not know what is.

The self-imposed convention of science that limits inquiry to objective observations of the world and generally seeks natural accounts for all phenomena is called *methodological naturalism*. We have also noted that methodological naturalism is often conflated with *metaphysical naturalism*, which assumes that reality itself is purely natural, that is, composed solely of material objects.

Methodological naturalism can still be applied without implying any dogmatic attachment to metaphysical naturalism. The thesis of this book is that the supernatural hypothesis of God is testable, verifiable, and falsifiable by the established methods of science. We can imagine all sorts of phenomena that, if observed by means of methodological naturalism, would suggest the possibility of some reality that is highly unlikely to be consistent with metaphysical naturalism.

For example, it could happen that a series of carefully controlled experiments provide independent, replicable, statistically significant evidence that distant, intercessory prayer of a specific kind, say, Catholic, cures certain illnesses while the prayers of other religious groups do not. It is difficult to imagine any plausible natural explanation for this hypothetical result.

IMPOSSIBLE GODS

Before proceeding with the scientific evidence bearing on the God hypothesis, let us make a quick review of those disproofs of God's existence that are based on philosophy. For a recent survey, see *The Non-Existence of God* by Nicholas Everitt.[12] Philosophers Michael Martin and Ricki Monnier have assembled a volume of essays on the logical arguments claiming to show the impossibility of gods with various attributes.[13] Here is how they classify these types of disproofs:

- definitional disproofs based on an inconsistency in the definition of God
- deductive evil disproofs based on the inconsistency between the existence of God who has certain attributes and the existence of evil
- doctrinal disproofs based on an inconsistency between the attributes of God and a particular religious doctrine, story, or teaching about God
- multiple-attribute disproofs based on an inconsistency between two or more divine attributes
- single-attribute disproofs based on an inconsistency within just one attribute

These disproofs merit greater credence than the claimed philosophical proofs of the existence of God, for the same reason scientists and philosophers give more credence to falsifications of scientific models than to the verifications. The logical disproofs seem inescapable, unless you change the rules of the game or, more commonly, change the definitions of the words being used in the argument.

In the following, formal statements for a sample of nonexistence arguments are listed, just to give the reader the flavor of the philosophical debate. They will not be discussed here since they are independent of the scientific arguments that form my main thesis; the conclusions of this book are in no way dependent on their validity. They are listed for completeness and for contrast with the scientific arguments. For the details, see the individual essays in the compilation by Martin and Monnier.[14]

The first two are examples of definitional disproofs:

An All-Virtuous Being Cannot Exist

1. God is (by definition) a being than which no greater being can be thought.
2. Greatness includes the greatness of virtue.
3. Therefore, God is a being than which no being could be more virtuous.
4. But virtue involves overcoming pains and danger.
5. Indeed, a being can only be properly said to be virtuous if it can suffer pain or be destroyed.
6. A God that can suffer pain or is destructible is not one than which no greater being can be thought.
7. For you can think of a greater being, one that is nonsuffering and indestructible.
8. Therefore, God does not exist.[15]

Worship and Moral Agency

1. If any being is God, he must be a fitting object of worship.
2. No being could possibly be a fitting object of worship, since worship requires the abandonment of one's role as an autonomous moral agent.
3. Therefore, there cannot be any being who is God.[16]

We have already briefly noted the problem of evil, and will be saying much more about it. For now, let us just indicate its formal statement:

The Problem of Evil

1. If God exists, then the attributes of God are consistent with the existence of evil.
2. The attributes of God are not consistent with the existence of evil.
3. Therefore, God does not and cannot exist.[17]

The following three are examples of multiple-attribute disproofs:

A Perfect Creator Cannot Exist

1. If God exists, then he is perfect.
2. If God exists, then he is the creator of the universe.
3. If a being is perfect, then whatever he creates must be perfect.
4. But the universe is not perfect.
5. Therefore, it is impossible for a perfect being to be the creator of the universe.
6. Hence, it is impossible for God to exist.[18]

A Transcendent Being Cannot Be Omnipresent

1. If God exists, then he is transcendent (i.e., outside space and time).
2. If God exists, he is omnipresent.
3. To be transcendent, a being cannot exist anywhere in space.
4. To be omnipresent, a being must exist everywhere in space.
5. Hence it is impossible for a transcendent being to be omnipresent.
6. Therefore, it is impossible for God to exist.[19]

A Personal Being Cannot Be Nonphysical

1. If God exists, then he is nonphysical.
2. If God exists, then he is a person (or a personal being).
3. A person (or personal being) needs to be physical.
4. Hence, it is impossible for God to exist.[20]

Finally, here is an example of a single-attribute disproof:

The Paradox of Omnipotence

1. Either God can create a stone that he cannot lift, or he cannot create a stone that he cannot lift.
2. If God can create a stone that he cannot lift, then he is not omnipotent.
3. If God cannot create a stone that he cannot lift, then he is not omnipotent.
4. Therefore, God is not omnipotent.[21]

The reader will undoubtedly see much in these bare formal statements that needs clarification; again I address you to the original essays for details and additional disproofs of this kind. Like most philosophical discussions, it mainly comes down to the meanings of words and assembling them into coherent, consistent statements. The philosophers who formulated these disproofs have been careful about defining the terms used, while those who dispute them will generally disagree with those definitions or the way they have been interpreted. As a result, the debate continues.

WAYS OUT

Ways out of purely logical arguments can always be found, simply by relaxing one or more of the premises or, as noted, one of the definitions. For example, assume God is not omnibenevolent. Indeed, the God of the more conservative elements of Judaism, Christianity, and Islam that take their scriptures literally can hardly be called omnibenevolent—or even very benevolent. No one reading the Bible or Qur'an literally can possibly regard the God described therein as all-good. We will see examples later, but for now the reader is invited to simply pick up an Old Testament or Qu'ran, open to a random page, and read for a while. It will not take you long to find an act or statement of God that you find inconsistent with your own concepts of what is good. And, as we will also see, much in the Gospel can hardly be called "good."

In any event, the scientific case is not limited to an omnibenevolent, omniscient, or omnipotent god.

The scientific method incorporates a means to adjudicate disputes that otherwise might run in circles, never converging as disputants on all sides of an issue continually redefine and refine their language. In science we are able to break out of this vicious cycle by calling upon empirical observations as the final judge. Of course, ways out of the scientific arguments can also be achieved by redefining God or by disputing the empirical facts. The reader will simply have to judge for herself whether the examples I present are convincing.

MODELS AND THEORIES

Science is not just a matter of making observations but also developing models to describe those observations. In fact, philosophers have pointed out that any observation or measurement we make in science depends on some model or theory. They assert that all observations are "theory-laden." For example, when we measure the time it takes for a particle to move from one point to another we must first assume a model in which particles are visualized as moving in space and time. The model must begin by defining space and time.

The use of models, which are simplified pictures of observations, is not limited to the professional practice of science. They are often used to deal with the ordinary problems of life. For example, we model the sun as an orb rising in the east and setting in the west. Travelers heading to the west can point themselves each day in the direction of the setting sun and, correcting for some northward or southward drift (depending on season), arrive safely at their destination. No additional elements to the model are needed—in particular, no metaphysics. The ancient Greeks viewed the sun as the gold-helmeted god Apollo, driving a golden chariot across the sky. The ancient Chinese thought it was a golden bird. Neither metaphysical model offers any additional aid to our travelers in their navigation. And, that lack of necessity in the absence of any other evidence testifies strongly for the nonexistence of such a god or golden bird.

While utilizing models is a normal process in everyday life, scientific models objectify and, whenever possible, quantify the procedure—thus providing a rational means for distinguishing between what works and what does not. Whenever possible, mathematics and logic are used as tools to enforce a consistency that is not always found in commonplace statements, which are formulated in the vernacular. For example, instead of saying that your blood pressure is probably high, a physician will measure it and give you two numbers, say, 130 over 100. Then he might prescribe some calculated amount of medication to bring the 100 down to 80.

Scientific instruments that enhance the power of our senses commonly yield quantifiable measurements, enabling scientists to deal with variables having numerical values upon which all observers can agree—within equally quantifiable measurement errors. While some sciences may deal with nonnumerical variables, physical models are almost always quantitative and the logical power of mathematics is put to great use in their utilization.

Most scientific models begin by defining their observables operationally, that is, by characterizing them in terms of a well-prescribed, repeatable measuring procedure. For example (as Einstein emphasized), time is defined as what you read on a clock. Temperature is what you read on a thermometer. Specific instruments are chosen as standards. A mathematical framework is then formulated that defines other variables as functions of the observables and postulates connections between these quantities.

The term *model* usually applies to the preliminary stages of a scientific process when considerable testing and further work still need to be done. The "theories" that arise from this effort are not the unsupported speculations that they are often accused of being by those unfamiliar with the scientific method or by those wishing to demean it. To be accepted into the ranks of scientific knowledge, theories must demonstrate their value by passing numerous, risky empirical tests and by showing themselves to be useful. Theories that fail these tests, or do not prove useful, are discarded.

In this book we will make frequent reference to the *standard models* of fundamental physics and cosmology. By now these have sufficiently advanced to the level where they can be honestly recognized as standard *theories*, although their prior designations as models continue to be used in the literature, presumably to maintain familiarity. I find it amusing and ironic that opponents of evolution think they are undermining it by calling it "just a theory."

The validity of the scientific method is justified by its immense success. However, we must recognize and acknowledge that scientific models and theories, no matter how well established, are still human contrivances and subject to change by future developments. This is in contrast to revelations from God, which should be true unconditionally and not subject to revision. Furthermore, the elements of scientific models, especially at the deepest level of quantum phenomena, need not correspond precisely to the elements of whatever "true reality" is out there beyond the signals we receive with our senses and instruments. We can never know when some new model will come along that surpasses the old one. We regard such a happening as the welcome progress of science rather than some disastrous revolution that tears down the whole prior edifice, rendering it worthless. For example, despite a common misunderstanding, the models of Newtonian mechanics were hardly rendered useless by the twin twentieth-century developments of relativity and quantum mechanics. Newtonian physics continue to find major application in contemporary science and technology. It is still what most students learn in physics classes and what most engineers and others use when they apply physics in their professions.

Perhaps quarks and electrons are not real, although they are part of the highly successful standard model of particle physics. We cannot say. But we can say, with high likelihood, that some of the elements of older models, such as the ether, are not part of the real world. And, while we cannot prove that every variety of god or spirit does not exist in a world beyond the senses, we have no more rational basis for including them than we have for assuming that the sun is a god driving a chariot across the sky. Furthermore, we can proceed to put our models to practical use without ever settling any metaphysical questions. Metaphysics has surprisingly little use and would not even be worth discussing if we did not have this great desire to understand ultimate reality as best as we can.

The ingredients of scientific models are not limited to those supported by direct observation. For example, the standard model of elementary particles and forces contains objects such as *quarks*, the presumed constituents of atomic nuclei, which have never been seen as free particles. In fact, the theory in its current form requires that they *not* be free. The observation of a free quark would falsify that aspect of the standard model, although nicely confirm the quark idea itself.

Indeed, the development of models in physics is often motivated by considerations of logical and mathematical beauty, such as symmetry principles. But they still must be tested against observations.

Astronomical models include black holes, which can only be observed indirectly. Cosmological models include dark matter and dark energy, which remain unidentified at this writing but are inferred from the data. The models currently used in modern physics, astronomy, and cosmology are solidly grounded on direct observations and have survived the most intensive empirical testing. By virtue of this success, they can be used to make inferences that are surely superior to speculations simply pulled out of thin air.

Physicists generally speak as if the unobserved elements of their models, such as quarks, are "real" particles. However, this is a metaphysical assumption that they have no way of verifying and, indeed, have no real need (or desire) to do so. The models of physics and their unobserved elements are human inventions and represent the best we can do in describing objective reality. When a model successfully describes a wide range of observations, we can be confident that the elements of those models have something to do with whatever reality is out there, but less confident that they constitute reality itself.

On the other hand, if a model does not work there is no basis to conclude that any unique element of that model is still part of reality. An example is the electromagnetic ether, which was discussed earlier.

Having read this, please do not assume that the doctrine of *postmodernism* is being promoted here. Science is decidedly not just another cultural narrative. The science referred to is called "Western science," which was developed originally by Europeans utilizing mathematical insights from India (the concept of "zero"), the Arab world (numerals, algebra), and other cultures. Peoples in all but the most primitive societies now utilize science. While we might consider science another "cultural narrative," it differs from other cultural narratives because of its superior power, utility, and universality.

MODELING GOD

Everyone involved in discourses on the existence of gods may be well advised to consider the approach outlined above. Like quarks, the gods are human inventions based on human concepts. Whether or not we can say if the gods people talk about have anything to do with whatever objective reality is out there depends on the empirical success of the models that are built around these hypothetical entities. Whatever a god's true nature, if one exists, a god model remains the best we can do in talking about that god.

If we accept this procedure, then we can eliminate a whole class of objections that are made to types of logical and scientific arguments formulated in this book. In these arguments, God is assumed to have certain attributes. The theologian may ask: how can we mere mortals know about the true nature of a god who lies beyond our sensibilities? The answer is that we do not need to know—just as physicists do not need to know the ultimate reality behind quarks. Physicists are satisfied that they have a model, which currently includes quarks, that agrees beautifully with the data. The quark model is empirically grounded. It represents the best we humans have been able to do thus far in describing whatever objective reality underlies nuclear and sub-nuclear observations. Whether quarks are real or not does not change this. Whether any of the objects of scientific models are real or not does not change the fact that those models have immense utility. This includes Newtonian physics, despite the further developments of twentieth-century relativity and quantum mechanics.

Analogously, if a particular god model successfully predicts empirical results that cannot be accounted for by any other known means, then we would be rational in tentatively concluding that the model describes some aspect of an objective reality without being forced to prove that god really is as described in the details of the model.

Still, any god model remains a human invention, formulated in terms of human qualities that we can comprehend, such as love and goodness. Indeed, the gods of ancient mythology—including the Judeo-Christian-Islamic God—are clearly models contrived by humans in terms people could understand. What is amazing is that in this sophisticated modern age so many still cling to primitive, archaic images from the childhood of humanity.

On the flip side, when a model is strongly falsified by the data, then those elements of the model that have been severely tested by observations should be rejected as not very likely to be representative of an objective reality.

The following example should illustrate this rather subtle concept. Observations of electromagnetic phenomena support a model of electromagnetism containing pointlike electric charges we can call *electric monopoles*. Examples include ions, atomic nuclei, electrons, and quarks. Symmetry arguments would lead you to include in the model point magnetic charges—*magnetic monopoles*.

Yet the simplest observed magnetic sources are described as *magnetic dipoles*—bar magnets that have north and south poles.

Electric dipoles such as hydrogen atoms, with a positive and a negative point charge separated in space, exist as well. But you can tear them apart into separate electric monopoles, such as an electron and a proton. On the other hand, if you cut away a piece of the north pole of a bar magnet, instead of getting a separate north and south monopole you get two dipoles—two bar magnets.

Despite these empirical facts, some theoretical basis exists for magnetic monopoles, and they have been searched for extensively with no success. The current standard model contains perhaps a single magnetic monopole in the visible universe, which has no effect on anything. That is, the model does include a magnetic monopole, but we can proceed to use our conventional electromagnetic theory, which contains no magnetic monopoles, for all practical applications.

Let us apply this same line of reasoning to God. When we show that a particular model of God fails to agree with the data, then people would not be very rational in using such a model as a guide to their religious and personal activities. While it remains possible that a god exists analogous to the lonely magnetic monopole, one who has no effect on anything, there is no point worshiping him. The gods we will consider are important elements of scientific models that can be empirically tested, such as by the successful consequences of prayer.

THE SCIENTIFIC GOD MODEL

So, let us now define a scientific God model, a *theory of God*. A supreme being is hypothesized to exist having the following attributes:

1. God is the creator and preserver of the universe.
2. God is the architect of the structure of the universe and the author of the laws of nature.

3. God steps in whenever he wishes to change the course of events, which may include violating his own laws as, for example, in response to human entreaties.
4. God is the creator and preserver of life and humanity, where human beings are special in relation to other life-forms.
5. God has endowed humans with immaterial, eternal souls that exist independent of their bodies and carry the essence of a person's character and selfhood.
6. God is the source of morality and other human values such as freedom, justice, and democracy.
7. God has revealed truths in scriptures and by communicating directly to select individuals throughout history.
8. God does not deliberately hide from any human being who is open to finding evidence for his presence.

Most of these attributes are traditionally associated with the Judeo-Christian-Islamic God, and many are shared by the gods of diverse religions. Note, however, that the traditional attributes of omniscience, omnipotence, and omnibenevolence—the 3O characteristics usually associated with the Judeo-Christian-Islamic God—have been omitted. Such a God is already ruled out by the arguments of logical inconsistency summarized above. While the 3Os will show up on occasion as supplementary attributes, they will rarely be needed. For example, the case against a creator god will apply to any such god, even an evil or imperfect one. Furthermore, as will be emphasized throughout, the God of the monotheistic scriptures—Old Testament or Hebrew Bible, New Testament, and Qur'an—is not omnibenevolent, and so not ruled out by logical inconsistency. The observable effects that such a God may be expected to have are still testable by the normal, objective processes of science.

THE GENERIC ARGUMENT

The scientific argument against the existence of God will be a modified form of the lack-of-evidence argument:

1. Hypothesize a God who plays an important role in the universe.
2. Assume that God has specific attributes that should provide objective evidence for his existence.
3. Look for such evidence with an open mind.
4. If such evidence is found, conclude that God *may* exist.
5. If such objective evidence is not found, conclude beyond a reasonable doubt that a God with these properties does *not* exist.

Recall that it is easier to falsify a hypothesis than verify one. The best we can do if the data support a particular god model is acknowledge that faith in such a God is rational. However, just as we should not use a failed physical model that does not work, it would be unwise for us to guide our lives by religions that worship any gods that fail to agree with the data.

NOTES

1. Theodore M. Drange, *Nonbelief and Evil: Two Arguments for the Nonexistence of God* (Amherst, NY: Prometheus Books, 1998), p. 41.

2. See also John L. Schellenberg, *Divine Hiddenness and Human Reason* (Ithaca, NY: Cornell University Press, 1993).

3. For a good example of data mining, see my discussion of the experiment by Elisabeth Targ and collaborators in Victor J. Stenger, *Has Science Found God? The Latest Results in the Search for Purpose in the Universe* (Amherst, NY: Prometheus Books, 2003), pp. 250–53.

4. Karl Popper, *The Logic of Scientific Discovery*, English ed.

(London: Hutchinson; New York: Basic Books, 1959). Originally published in German (Vienna: Springer Verlag, 1934).

5. Rudolf Carnap, "Testability and Meaning," *Philosophy of Science* B 3 (1936): 19–21; B 4 (1937): 1–40.

6. Philip J. Kitcher, *Abusing Science: The Case Against Creationism* (Cambridge, MA: MIT Press, 1982). Note that the author was refuting the common creationist claim that evolution is not science because it is not falsifiable. Kitcher need not have bothered. Evolution is eminently falsifiable, as we show in chapter 3.

7. I discuss several examples in Victor J. Stenger, *Physics and Psychics: The Search for a World beyond the Senses* (Amherst, NY: Prometheus Books, 1990).

8. Karl Popper, "Metaphysics and Criticizability," in *Popper Selections*, ed. David Miller (Princeton, NJ: Princeton University Press, 1985), p. 214. Originally published in 1958.

9. Ibid.

10. National Academy of Sciences, *Teaching About Evolution and the Nature of Science* (Washington, DC: National Academy of Sciences, 1998), p. 58. Online at http://www.nap.edu/catalog/5787.html (accessed March 5, 2006).

11. Phillip E. Johnson, *Evolution as Dogma: The Establishment of Naturalism* (Dallas, TX: Haughton Publishing Co., 1990); *Darwin on Trial* (Downers Grove, IL: InterVarsity Press, 1991); *Reason in the Balance: The Case Against Naturalism in Science, Law, and Education* (Downers Grove, IL: InterVarsity Press, 1995); *Defeating Darwinism by Opening Minds* (Downers Grove, IL: InterVarsity Press, 1997); *The Wedge of Truth: Splitting the Foundations of Naturalism* (Downers Grove, IL: InterVarsity Press, 2001).

12. Nicholas Everitt, *The Non-Existence of God* (London, New York: Routledge, 2004).

13. Michael Martin and Ricki Monnier, eds., *The Impossibility of God* (Amherst, NY: Prometheus Books, 2003).

14. Ibid.

15. Douglas Walton, "Can an Ancient Argument of Carneades on Cardinal Virtues and Divine Attributes Be Used to Disprove the Existence of God?" *Philo* 2, no. 2 (1999): 5–13; reprinted in Martin and Monnier, *The Impossibility of God*, pp. 35–44.

16. James Rachels, "God and Moral Autonomy," in *Can Ethics Provide Answers? And Other Essays in Moral Philosophy* (New York: Rowman & Littlefield, 1997), pp. 109–23; reprinted in Martin and Monnier, *The Impossibility of God*, pp. 45–58.

17. Martin and Monnier, *The Impossibility of God*, p. 59.

18. Theodore M. Drange, "Incompatible-Properties Arguments—A Survey," *Philo* 1, no. 2 (1998): 49–60; in Martin and Monnier, *The Impossibility of God*, pp. 185–97.

19. Ibid.

20. Ibid.

21. J. L. Cowen, "The Paradox of Omnipotence Revisited," *Canadian Journal of Philosophy* 3, no. 3 (March 1974): 435–45; reprinted in Martin and Monnier, *The Impossibility of God*, p. 337.

Chapter 2

THE ILLUSION OF DESIGN

Look round this universe. What an immense profusion of beings ani-
mated and organized, sensible and active! . . . But inspect a little more
narrowly these living existences. . . . How hostile and destructive to
each other! How insufficient all of them for their own happiness!
                                                                    —David Hume

PALEY'S WATCH

Perhaps no argument is heard more frequently in support of
the existence of God than the *argument from design*. It represents the most common form of the God of the gaps argument:
the universe and, in particular, living organisms on Earth are said
to be simply too complex to have arisen by any conceivable natural mechanism.

Before the age of science, religious belief was based on faith, cultural tradition, and a confidence in the revealed truth in the scriptures and teachings of holy men and women specially selected by God. As science began to erode these beliefs by showing that many of the traditional teachings, such as that of a flat Earth at rest at the center of a firmament of stars and planets were simply wrong, people began to look to science itself for evidence of a supreme being that did not depend on any assumptions about the literal truth of the Bible or divine revelation.

The notion that the observation of nature alone provides evidence for the existence of God has a long history. It received perhaps its most brilliant exposition in the work of Anglican archdeacon William Paley (d. 1805). In his *Natural Theology or Evidences of the Existence and Attributes of the Deity Collected from the Appearance of Nature*, first published in 1802,[1] Paley wrote about finding both a stone and a watch while crossing a heath. While the stone would be regarded as a simple part of nature, no one would question that the watch is an artifice, designed for the purpose of telling time. Paley then alleged that objects of nature, such as the human eye, give every indication of being contrivances.

Paley's argument continues to be used down to the present day. Just a few weeks before writing these words, two Jehovah's Witnesses came to my door. When I politely expressed my skepticism, one began, "Suppose you found a watch . . ." Design arguments never die; nor do they fade away.

Sophisticated modern forms of the argument from design are found in the current movement called *intelligent design*, which asserts that many biological systems are far too complex to have arisen naturally. Also classifying as an argument from design is the contemporary claim that the laws and constants of physics are "fine-tuned" so that the universe is able to contain life. This is commonly but misleadingly called the *anthropic principle*. Believers also often ask how the universe itself can have appeared, why there is something rather than nothing, how the laws of nature and human reason could possibly have arisen—all without the action of a supreme being who transcends the world of space, time, and matter. In this chapter and those that follow, we will see what science has to say about these questions.

DARWINISM

When Charles Darwin (d. 1882) entered Cambridge University in 1827 to study for the clergy, he was assigned to the same rooms in Christ's College occupied by William Paley seventy years earlier.[2] By that time, the syllabus included the study of Paley's works and Darwin was deeply impressed. He remarked that he could have written out the whole of Paley's 1794 treatise, *A View of the Evidences of Christianity*, and that *Natural Theology* "gave me as much delight as did Euclid."[3]

Yet it would be Darwin who provided the answer to Paley and produced the most profound challenge to religious belief since Copernicus removed Earth from the center of the universe. Darwin's discovery caused him great, personal grief and serves as an exemplar of a scientist following the evidence wherever its leads and whatever the consequences.

Although the idea of evolution had been around for a while, Darwin's grandfather Erasmus Darwin being a notable proponent, no one had recognized the mechanism involved. That mechanism, proposed by Darwin in 1859 in *The Origin of Species*[4] and independently by Alfred Russel Wallace,[5] was *natural selection* by which organisms accumulate changes that enable them to survive and have progeny that maintain those features. Darwin had actually held back publishing for twenty years until Wallace wrote him with his ideas and forced him to go public. Darwin's work was by far the more comprehensive and deserved the greater recognition it received.

Today we understand the process of natural selection in terms of the genetic information carried in the DNA of cells and how it is modified by random mutations. It is not my purpose here to give yet another exposition of evolution. Darwin's theory, updated by the many developments since his time, resides at the foundation of modern biology. Evolution by natural selection is accepted as an observed fact by the great majority of biologists and scientists in related fields, and is utilized in every aspect of modern life science including medicine. In terms of the same strict standards of empirical evidence that apply in all the natural sciences, Darwinian evolution is a well-established theory that has passed many critical tests.

A common argument made by opponents of evolution is that it is not a "true" scientific theory, like electromagnetism or thermodynamics. They wrongly claim that evolution does not make predictions that can be tested and is thus not falsifiable. In fact, evolution is eminently predictive and falsifiable.

Darwin specifically predicted that recognizable human ancestors would be found in Africa. Many now have been. Evolutionary theory predicted that the use of antiviral or antibacterial agents would result in the emergence of resistant strains. This principle is, of course, a mainstay of contemporary medicine. Paleontologists correctly predicted that species showing the evolution from fish to amphibian would be found in Devonian strata.

This example, among many, refutes the frequently heard creationist claim that "transitional forms" (presumably meaning transitional species) do not exist. Paleontologists had expected to find transitions from land-based mammals to whales for years. In the past decade, science journals, as well as the media, have been full of these finds. A simple Internet search will yield hundreds of examples of transitional species.

The failure of many of these predictions would have falsified evolution. They did not fail. It is a trivial exercise to think of other ways to falsify evolution. For example, evolution would be falsified if we were to find bona fide remains of organisms out of place in the fossil record. Suppose mammals (horses, humans, or hippos) were found in the Paleozoic strata associated with trilobites, crinoids, and extinct corals. This would show that there was no evolutionary process. But we do not find any such inconsistencies.

My favorite example is over a hundred years old. Shortly after its publication in the nineteenth century, the theory of evolution was challenged by the famous physicist William Thomson, Lord Kelvin, whose thermodynamic calculations gave an age for Earth that was much too short for natural selection to operate. Darwin regarded this as the most serious challenge to his theory.

However, at the time, nuclear energy was unknown. When this new form of energy was discovered early in the twentieth century, Kelvin and other physicists quickly realized that the energy released by nuclear reactions at the center of the sun would be very efficient, allowing the sun and other stars to last billions of years as a stable energy source. In fact, evolution can be said to have predicted the existence of such an energy source! When he learned of nuclear energy, Kelvin graciously withdrew his objection to evolution.

As we will find several times in this book, some scientific arguments for the existence of God once had considerable force, and it was not until recently—within the last century—that accumulated knowledge not only eliminated these lines of reasoning but also turned many of them on their heads to support the case against God. These examples amply refute the claim that science has nothing to say about God. One can imagine endless scenarios by which observations of the universe and life on Earth might confirm God's existence; we will mention just a few in this book.

The discovery of human ancestors, the DNA and anatomical connections between humans and other animals (and even plants), and the use of animals in medical research falsify the hypothesis of a God who created humans as a distinct life-form. The fossil record, the existence of transitional species, and the actual observation of evolution in the laboratory falsify the hypothesis of a God who created separate "kinds" or species of life-forms at one time in history and left them unchanged since. It might have been otherwise.

Many believers see no conflict between evolution and their faith. After all, God can do anything he wants. If he wanted to create life by means of evolution, then that's what he did. However, other believers have good reason to regard evolution as threatening to their own faith in the purposeful, divine creation of human life.[6] Evolution implies humanity was an accident and not the special creature of traditional doctrine. Many find this unacceptable and conclude, despite the evidence, that evolution must be wrong.

However, if we are to rely on science as the arbiter of knowledge rather than ancient superstitions, the opposite conclusion is warranted. Evolution removes the need to introduce God at any step in the process of the development of life from the simplest earlier forms. It does not explain the origin of life, so this gap still remains. This is insufficient to maintain consistency for some believers, especially since evolution is in deep disagreement with the biblical narrative of simultaneously created immutable forms. Furthermore, we have no reason to conclude that life itself could not have had a purely material origin.

THE CREATIONISTS

While a continuum of creationist views from extreme to moderate continues to be heard, we can still identify a few dominant strains. Let us look at the recent history. According to Ronald Numbers, author of the definitive early history *The Creationists*, the term *creationism* did not originally apply to all forms of antievolution.[7] Opponents of evolution were not always committed to the same, unified view of creation. However, by the 1920s, the biblical creation story became the standard alternative to evolution in the United States and the creationist movement its champion.

In that decade, Christian fundamentalists in the United States took over the front line of the battle. Under their influence, three states—Tennessee, Mississippi, and Arkansas—made the teaching of evolution a crime. Oklahoma prohibited textbooks promoting evolution, and Florida condemned the teaching of Darwinism as "subversive."

In 1925 biology teacher John Scopes was brought to court in Dayton, Tennessee, for teaching evolution. This led to the sensational "Monkey Trial," with Clarence Darrow for the defense pitted against three-time losing Democratic presidential candidate William Jennings Bryan for the prosecution. Although Scopes was convicted (later overturned on appeal), the trial is still widely regarded as a public relations triumph for the Darwinians, as somewhat inaccurately depicted in the play and film *Inherit the Wind*.

A new strain of creationism appeared in 1961 with the publication of *The Genesis Flood* by theologian John C. Whitcomb Jr. and hydraulic engineer Henry M. Morris,[8] who were strongly influenced by earlier efforts by Seventh-day Adventist leader George McCready Price. The authors argued that science was compatible with Genesis, and although their scientific claims were not credible, conservative Christians sat up and took notice—recognizing a new strategy for combating hated Darwinism. Around 1970 Morris founded the Institute for Creation Science, which then led a movement to have the new "creation science" presented in public-school science classrooms. Biochemist Duane Gish traveled the country on behalf of the institute, giving talks and ambushing naive biologists in debates before huge, receptive audiences of churchgoers. Arkansas and Louisiana passed laws mandating the teaching of creation science alongside evolution.

In 1982 a federal judge in Arkansas tossed out the law in that state, declaring creation science to be religion and not science.[9] In 1987 the Supreme Court ruled the Louisiana law unconstitutional.

About this time, creation science speciated into two main branches, one holding to the more literal biblical picture of a

young Earth and another that attempts to use sophisticated arguments that appear, at least to the untutored eye, more consistent
with established science. The second group has developed a new
stealth creationism called *intelligent design*, which has the
common shorthand, "ID."

THE WEDGE OF INTELLIGENT DESIGN

Learning from the mistakes of the creation scientists, proponents
of ID downplay their religious motives in a so far not very successful attempt to steer clear of the constitutional issue. They also
have avoided the more egregious scientific errors of the young-
Earth creationists, and present this new form of creationism as
"pure science." They claimed that design in nature can be scientifically demonstrated and that the complexity of nature can be
proved not to have arisen by natural processes alone.[10]

In *Creationism's Trojan Horse: The Wedge of Intelligent Design*,
philosopher Barbara Forrest and biologist Paul Gross detail the
story of how the new creationism is fed and watered by a well-
funded conservative Christian organization called the Discovery
Institute.[11] The goals of this organization, documented by Forrest
and Gross, are to "defeat scientific materialism and its destructive
moral, cultural, and political legacies" and to "renew" science
and culture along evangelical Christian lines.

BEHE'S IRREDUCIBLE COMPLEXITY

None of the claims of intelligent design proponents, especially
the work of its primary theorists, biochemist Michael Behe and
theologian William Dembski, have stood up under scientific
scrutiny. Numerous books and articles have refuted their positions in great detail.[12] Not only have their arguments been shown

to be flawed, but also in several instances the factual claims on which they rest have been proven false. None of their work has been published in respected scientific journals.[13]

Behe's fame rests on his 1996 popular-level book, *Darwin's Black Box: The Biochemical Challenge to Evolution*.[14] There he introduced the notion of *irreducible complexity*, which occurs when a system is reduced to several parts and can no longer function when any of the parts is removed. Behe argued that the individual parts could not have evolved by natural selection since they no longer have any function on which selection can operate.

Thoroughly refuting Behe's argument, evolutionary biologists have listed many examples in nature where an organic system changes functions as the system evolves.[15] They have provided plausible natural mechanisms for every example Behe presents, many of which were well known (except to Behe) before Behe ever sat down to write.

The manner in which the parts of living systems change function over the course of evolution is one of those well-established facts of evolution that Behe and other proponents of intelligent design choose to ignore. Biological parts often evolve by natural selection by virtue of one function, and then gradually adapt to other functions as the larger system evolves.

Many examples of organs and biological structures that are understood to have arisen from the modification of preexisting structure rather than the elegance of careful engineering can be found in the biological literature. Paleontologist Stephen Jay Gould made this point in his wonderful example of the panda's thumb.[16] The panda appears to have six fingers, but its opposing "thumb" is not a finger at all but a bone in its wrist that has been enlarged to form a stubby protuberance handy for holding a stalk of bamboo shoots, the panda's only food.

Behe is a biochemist, not an evolutionary biologist, and was unaware when he wrote his book that the mechanisms for the evolution of "irreducibly complex" systems were already discussed six decades earlier by the Nobel Prize winner Hermann Joseph Muller and have been common knowledge in the field since then.[17] Behe cannot even be forgiven for simply falling into the God of the gaps trap. He did not even find a gap.

THE EYE

Let us look at the frequent example used by creationists since Paley: the human eye. In *The Blind Watchmaker*, which was primarily a contemporary evolution scientist's response to William Paley, zoologist Richard Dawkins pointed out that the eye in all vertebrates is wired backward, with the wires from each light-gathering unit sticking out on the side nearest the light and traveling over the surface of the retina where it passes through a hole, the "blind spot," to join to the optic nerve.[18] Other animals, such as the octopodes and squids, have their eyes wired more rationally.

This is often presented as an example of apparent "poor design." However, biologist (and devout Catholic) Kenneth Miller does not think this is a fair designation, since the arrangement still works pretty well. He has shown how the wiring of the vertebrate is nicely described by evolution.[19] The retina of the eye evolved as a modification of the outer layer of the brain that gradually developed light sensitivity. The eye is neither poorly nor well designed. It is simply not designed.

Eyes provide such obvious survival value that they developed at least forty times *independently* in the course of evolution.[20] Neuroscience has identified eight different optical solutions for collecting and focusing light, although all share similarities at the molecular and genetic levels.[21] The physics and chemistry are the same; few ways exist for detecting photons. But, because of the important role of chance and local environment in the evolution of complex systems, different solutions to the problem were uncovered by random sampling of the varied paths allowed by evolution. In short, the structures of eyes look as they might be expected to look if they developed from purely material and mindless processes— chance plus natural selection—as these processes explore the space of possible survival solutions.

DEMBSKI'S INFORMATION

While to this date Behe has written one book, his Discovery Institute colleague William Dembski has been highly prolific, with several books and many articles on intelligent design.[22] Dembski claims that design in nature is mathematically demonstrable. Since his arguments are couched in highly and often ambiguous technical language, they require a certain expertise to understand and evaluate. Fortunately, many experts have taken the trouble to carefully examine Dembski's work. Almost universally they show it to be deeply flawed.[23] I will just mention here one example where Dembski, like Behe, makes statements that are provably wrong.

In his popular book *Intelligent Design: The Bridge between Science and Theology* (no hiding the religious motive here), Dembski asserts, "Chance and law working in tandem cannot generate information."[24] He calls this the *Law of Conservation of Information*.

In *Has Science Found God?* I disproved this "law" by simply and trivially showing that the quantitative definition of information, as used conventionally and, somewhat obscurely, by Dembski is equivalent to negative entropy.[25] Entropy, which is the quantitative measure of disorder in physics (hence information being related to negative entropy, or order), is not a conserved quantity like energy. In fact, the entropy of an "open" system (one that interacts with its environment by exchanging energy) can either increase or decrease. Certainly living systems on Earth are open systems. Indeed, a living organism is kept away from thermodynamic equilibrium by its use of sources of outside energy to maintain order.

THE POLITICAL BATTLE TODAY

While at this writing intelligent design continues to gain adherents among those believers who cannot reconcile Darwinian natural selection with their faith, scientists of many faiths and scientists of no faith have agreed overwhelmingly that intelligent design has not made its case scientifically. All the major scientific societies in the United States have issued statements supporting evolution and rejecting intelligent design. Behe's own department at Lehigh University has put it as well as any:

> The faculty in the Department of Biological Sciences is committed to the highest standards of scientific integrity and academic function. This commitment carries with it unwavering support for academic freedom and the free exchange of ideas. It also demands the utmost respect for the scientific method, integrity in the conduct of research, and recognition that the validity of any scientific model comes only as a result of rational hypothesis testing, sound experimentation, and findings that can be replicated by others.
>
> The department faculty, then, are unequivocal in their support of evolutionary theory, which has its roots in the seminal work of Charles Darwin and has been supported by findings accumulated over 140 years. The sole dissenter from this position, Prof. Michael Behe, is a well-known proponent of "intelligent design." While we respect Prof. Behe's right to express his views, they are his alone and are in no way endorsed by the department. It is our collective position that intelligent design has no basis in science, has not been tested experimentally, and should not be regarded as scientific.[26]

Amid faculty protests, Dembski has left Baylor University, the largest Baptist university in the world, for the Southern Baptist Theological Seminary.[27] Many scholars at Baylor and other Christian universities have come to realize that intelligent design does not provide respectable support for their religious beliefs.[28]

The battle over intelligent design, which is fought in the political arena rather than in scientific venues, is producing its share of litigation.[29] In a court case that attracted world attention in December 2005, a federal court in Dover, Pennsylvania, determined that intelligent design was motivated by religion and thus presenting it in science classes in public schools is unconstitutional.[30] This would seem to signal the death knell for intelligent design except for a subtle point that has escaped the notice of most of the scientific community and others that support evolution.

In the Dover trial Judge John E. Jones III ruled that teaching intelligent design (ID) in public-school science classes is an unconstitutional violation of church and state. This case mirrored *McLean v. Arkansas*, described above.

In both trials, the presiding federal judges went further than was necessary in making their rulings. Not only did the jurists rule creation science and ID as unconstitutional entanglements of government with religion, which would have been sufficient to decide each case (as Judge Jones admitted in his decision), but they also labeled them as not science. In doing so, they were forced to define science—something on which neither scientists nor philosophers have been able to reach a consensus.

In Arkansas, Judge William R. Overton relied mainly on the testimony of philosopher Michael Ruse and defined science as follows:[31]

(1) It is guided by natural law;
(2) It has to be explained by reference to natural law;
(3) It is testable against the empirical world;
(4) Its conclusions are tentative, that is, are not necessarily the final word;
(5) It is falsifiable.

The eminent philosopher Larry Laudan, my colleague at the University of Hawaii at the time, had worked for years on the so-called demarcation problem, how to draw a line between science and nonscience. When the Arkansas decision was announced, Laudan objected strenuously. He pointed out that creation science is in fact testable, tentative, and falsifiable. For example, it predicts a young Earth and other geological facts that have, in fact, been falsified. Falsified science can still be science, just wrong science. Laudan warned that the Arkansas decision would come back to haunt science by "perpetuating and canonizing a false stereotype on what science is and how it works."[32]

Coming up to date, we similarly find that intelligent design is testable, tentative, and falsifiable. As described above, the claims of primary design theorists William Dembski and Michael Behe have been thoroughly refuted and in some cases falsified.

I am not quibbling with the ruling that ID, as practiced by the Dover Board of Education, represented an unconstitutional attempt to promote a sectarian view of creation under the guise of science. And I also agree that ID has all the markings of pseudoscience rather than genuine science.

Judge Jones relied on the Arkansas precedent and witnesses from both sides who testified that for ID to be considered science, the ground rules of science would have to be broadened to allow the consideration of supernatural forces. This position was both unwise and incorrect, for reasons I discussed in chapter 1. It is unwise because it plays into the hands of those who accuse science of dogmatism in refusing to consider the possibility on nonnatural elements at work in the universe. It is incorrect because science is not forbidden from considering supernatural causes. Furthermore, some reputable scientists are doing just that.

SELF-ORGANIZATION

Proponents of intelligent design often point to a statement by "400 scientists" that is purported to demonstrate their support for intelligent design. Let me quote the exact statement: "We are skeptical of claims for the ability of random mutation and natural selection to account for the complexity of life. Careful examination of the evidence for Darwinian theory should be encouraged."[33]

Note that "intelligent design" does not appear in the statement. In fact, it is rather a mild expression of skepticism, always a reasonable scientific attitude, and a gratuitous call for careful examination of the evidence for Darwin's theory—unnecessary because this has been the rule in evolution science since Darwin's voyage on the *Beagle*. Indeed, Darwin's work still serves as an exemplar of the best in empirical and theoretical science, and is one of the most strenuously tested.

Nevertheless, there may indeed be more to the mechanism of evolution than random mutation and natural selection. It simply isn't intelligent design. Complex material systems exhibit a purely natural process called *self-organization* and this appears to occur in both living and nonliving systems.

In his beautifully illustrated book *The Self-Made Tapestry*, Philip Ball gives many examples of pattern formation in nature that should provide a strong antidote for those who still labor under the delusion that mindless natural processes are unable to account for the complex world we see around us.[34] The fact that many patterns observed in biological systems are also present in nonliving systems and can be understood in terms of elementary, reductionist physics also should provide an antidote for those who still labor under the delusion that special *holistic* or *nonreductive* processes are needed to account for the complexity of life. Simplicity easily begets complexity in the world of locally interacting particles.[35] The whole is the sum of its parts.

One remarkable observation, for example, is the frequent

appearance in nature of the *Fibonacci sequence* of numbers. This is the set of numbers in which each entry is the sum of the preceding two: 0, 1, 1, 2, 3, 5, 8, 13, 21, 34, 55, . . . The number of petals on many flowers is a Fibonacci number. Buttercups have five petals, marigolds have thirteen, and asters have twenty-one.

Dembski has attempted to argue that the appearance of what he calls *complex specified information* is evidence for "intelligent design" in the universe. He claims that simple natural processes are incapable of producing complex specified information.[36] In his 1999 book, *Intelligent Design*, Dembski gives an example of the type of complex specified information that, when observed in nature, would in his view provide evidence for an intelligent source of that information. He refers to the film *Contact*, based on the novel of the same name by famed astronomer Carl Sagan.[37]

In the film, an extraterrestrial signal is observed by astronomers and interpreted as the sequence of prime numbers from 2 to 101. The astronomers in the story take this as evidence for an extraterrestrial intelligence. Dembski argues that many living things on Earth exhibit this kind of complex specified information that can only be produced by extraterrestrial, or perhaps extra-universal intelligence.

But Dembski does not have to wait for signals from outer space to provide an interesting mathematical sequence. He can walk out into his garden and count the petals on flowers. He will find that most contain "complex specified information" that comes from purely natural processes.

One example, given by Ball, is the double spiral pattern that is commonly found in nature. In 80 percent of plant species, leaves spiral up the stem, each separated from the one below by a constant angle turn.[38] A double spiral pattern, twisting in opposite directions, is seen when viewed from above. This double spiral pattern is also seen in the florets of flower heads such as the sunflower (see fig. 2.1) and the leaflets in a pinecone.

It might be thought that some biological process, perhaps

associated with Darwinian evolution, is taking place. However, it turns out to be simple physics—the minimization of potential energy.

In 1992 Stéphanie Douady and Yves Couder placed tiny droplets of magnetic fluid on a film of oil. They applied a vertical magnetic field that polarized the droplets and caused them to repel one another. Another field was applied along the periphery that pulled the droplets to the edge. They observed the droplets arrange themselves in a double spiral, thus demonstrating that the mechanism for spiral formation is physical rather than uniquely biological.[39]

Several computer simulations have reproduced this result. However, I decided to try one myself that made as few assumptions as possible. I started with an electrically charged particle, such as an electron, and added more particles one at a time in rings of increasing radius from the central particle. I chose a particle location in each ring as the position for which the electric potential energy for a particle in that ring is minimum. The result is shown in figure 2.2. We see that the double spiral pattern is reproduced. Please note that this pattern was not built into the algorithm used, which involved only minimizing the total poten-

Fig. 2.1.
A sunflower showing the double spiral pattern of florets in the flower head. Photograph by John Stone.

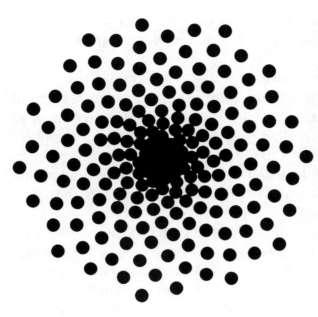

Fig. 2.2. The distribution of charged particles that minimizes potential energy. The double spiral pattern commonly seen in plants is reproduced.

tial energy, where the potential energy surrounding a point particle is spherically symmetric.

With this simple computer program, I have demonstrated the process called *spontaneous symmetry breaking*, whereby the symmetry of a system is broken naturally, that is, without being forced on the system by some asymmetric mechanism. We will see the importance of spontaneous symmetry breaking when we talk about the formation of structure in the universe in the absence of design.

Biologist Stuart Kauffman has long argued that self-organization plays a larger role in the evolution of life than previously thought, that blind natural selection is not sufficient.[40] He proposes that life originated by a chemical process known as *catalytic closure* and visualizes a network of interlinked chemical reactions becoming self-sustaining. Although Kauffman seems to imply that self-organization is some new, holistic law of nature, in fact nothing is needed besides basic, purely reductionistic physics and chemistry.

The origin of life itself is not accounted for by the theory of evolution. Some prebiological process such as self-organization must have been involved. This is a current gap in scientific knowledge, but plausible natural mechanisms such as Kauffman's are sufficient to keep God out of the picture.

SIMPLE RULES

In recent years, with the aid of computer simulations, we have begun to understand how simple systems can self-organize themselves into highly complex patterns that, at least superficially, resemble those seen in the world around us.[41] Usually these demonstrations start by assuming a few simple rules and then programming a computer to follow those rules. Some imagine they see a "law of increasing complexity" in which simple material systems become complex by self-organization.[42] I see no evidence for this, just the workings of well-known laws of particle mechanics applied to systems of many particles. In any case, such a law, if it exists, has nothing to do with whether the systems are living or nonliving.

The computer has made it possible for scientists to study many examples of complexity arising from simplicity. These are perhaps most easily demonstrated in what are called *cellular automata*, which were used by mathematician John von Neumann as an example of systems that can reproduce themselves. While cellular automata can be studied in any number of dimensions, they are easiest to understand in terms of a two-dimensional grid such as a piece of graph paper. You basically fill in a square on the grid based on a rule that asks whether or not certain of its adjoining squares are filled in. Note again that this is a purely "local" process, with no reference to cells that do not touch the cell in question.

Self-reproduction with cellular automata can be illustrated by a simple rule introduced by physicist Edward Fredkin in the 1960s.[43] Fill in a cell, that is, turn it "on," if and only if an odd number of the four nondiagonal neighbors (top, bottom, left, right) are on. Repeat this process on any initial pattern of cells, and that pattern will produce four copies of itself every four cycles (see fig. 2.3).

In a recently published, controversial tome called *A New Kind of Science*, physicist Stephen Wolfram has produced an enormous compilation of cellular automata.[44] Beyond these examples, Wolfram claims he has uncovered a "new kind of science" in which the universe itself operates like a digital computer. While he has presented some new proposals and numerous new examples, the original idea of a digital universe is usually attributed to Fredkin.[45] Whoever deserves the credit, it remains to be seen if this is a new science, since all that has been done so far are computer explorations of cellular automata with no connection to the real world yet established by predictions that can be tested empirically.

For my purposes here, suffice it to say that complex systems do not need complex rules in order to evolve from simple origins. They can do so with simple rules and no new physics. The grandiose claims one often hears in the literature about new

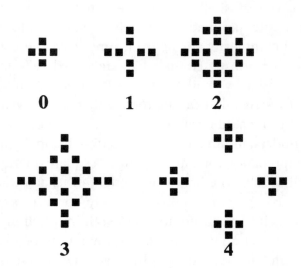

Fig. 2.3. Fredkin's self-reproducing cellular automaton.
The pattern at 0 produces four copies of itself in four steps.

holistic principles emerging from these processes are without foundation. It follows that no complex rule maker of infinite intelligence is implied by the existence of complex systems in nature. Since all we need are simple rules, then, at most, a simple rule maker of limited intelligence is required.

DEFINING DESIGN

Note that to make an argument *from* design assumes a priori that design exists. Philosopher Nicholas Everitt suggests that better terms might be the argument *from* order, or, the argument *to* design.[46] We will see that the evidence points firmly to the absence of design. And, if one of the attributes of God is that he designed the universe with at least one of his purposes being the existence of the complex structures we identify as life, with a special role for human life, then the failure to observe such design provides us with empirical grounds for concluding that a God with this attribute does not exist.

Some authors use the term "design" to refer to any structure of atoms and molecules that exhibits some pattern or purpose. Indeed, many are inconsistent in their usage and definition of the term "design."[47] In order to avoid any confusion on this matter, we will use *design* to refer to the act of an agent, be she divine or human, stupid or intelligent, to draw a blueprint—so to speak— of some artifact that is later assembled from that plan.

The assembly process in some cases might require high intelligence, as the Wright brothers demonstrated at every step when they built a flying machine in their bicycle shop. Or, the assembly can be relatively mindless, as on a modern automated production line—unless you want to argue that the computers running the process are pretty smart themselves. Indeed, many use the methods of "artificial intelligence." In any case, the assembly is unimportant unless the claim is being made that the assembly itself is a miracle. Since that is not normally an issue, what matters is the initial plan—a purpose that is either built into the contrivance from the beginning or not. In the example of the spiral discussed above, the broken symmetry of the spiral was not introduced by the programmer, me, on purpose.

Now, we must be careful not to confuse a preexisting purpose with mere utility or function. A stone can be used to break a window; however, the stone was not designed for that purpose. A salt crystal has a structure. But that structure was not contrived so that food would taste better when sprinkled with salt.

Similarly, all living organisms have many parts serving functions that are crucial for the survival of the organism. The question is: did an intelligent agent design that part for its present purpose, or did that function evolve by a combination of accident and the mechanisms of natural selection? In examining evidence for or against design in the world, we should look at whether the system being studied shows any sign of preexisting purpose or plan, or whether it can be seen to have evolved mindlessly by natural selection in response to the needs of survival or other purely physical mechanisms such as self-organization.

BAD DESIGN

As mentioned, Paley drew an analogy between different parts of the human body and an exquisitely designed watch. In such a watch, every part—the balance, escape wheel, jewel, mainspring, and the rest—is carefully constructed to serve its specific functions as efficiently as possible. The parts can always be improved upon, but not by much if the original work was by an expert craftsperson. Watches and all the many devices of human design have very few wasted parts.

Some evolutionists have tried to counter the Paley claim with what might be called the *argument from bad design*, pointing out all the ways that a competent engineer could improve upon what nature has given us.

The parts of the human body hardly resemble a watch. In an article in *Scientific American* titled "If Humans Were Built to Last," S. Jay Olshansky, Bruce Carnes, and Robert N. Butler have looked at flaws in the human body and shown how an engineer might have fixed them to enable us to live a hundred years or more in better health.[48] They trace our physical defects to the Rube Goldberg way evolution cobbles together new features by tinkering with existing ones. Natural selection does not seek out perfection or endless good health. The body has to live only long enough to reproduce and raise young. Species survival does not require that individuals survive long after reproducing. We humans do, albeit with decreasing vitality, because human evolution resulted in offspring that require years to mature and grandparents with enough years remaining to help in their upbringing. Speaking as a grandfather, thank you, evolution!

Let me list some of the flaws the *Scientific American* authors detect in the human machine that point away from any kind of near-perfection in design. Our bones lose minerals after age thirty, making them susceptible to fracture and osteoporosis. Our rib cage does not fully enclose and protect most internal organs. Our muscles atrophy. Our leg veins become enlarged and twisted, leading to varicose veins. Our joints wear out as their lubricants thin. Our retinas are prone to detachment. The male prostate enlarges, squeezing and obstructing urine flow.

Olshansky, Carnes, and Butler show what a properly designed human would be like. She would have bigger ears, rewired eyes, a curved neck, a forward-tilting torso, shorter limbs and stature, extra padding around joints, extra muscles and fat, thicker spinal disks, a reversed knee joint, and more. But she would not be very pretty by our present standards.

Despite their shortcomings, the various parts of the human body and those of other species do their jobs—even if those jobs were not part of any original plan. As discussed earlier, biologist Kenneth Miller argues persuasively that the eye serves us well and the inside-out nature of the vertebrate eye is nicely described by evolution.

NOWHERE EVIDENT

Richard Dawkins subtitled *The Blind Watchmaker* "Why the Evidence of Evolution Reveals a Universe without Design."[49] However, not just biological data but, as we will see in future chapters, the whole realm of scientific observations lead to the same conclusion: the universe does not look designed.

Estimates of the number of biological species on Earth range as high as one hundred million. Species on the order of ten or a hundred times this number once lived and have become extinct. Without getting into the current situation, where scientists and environmentalists fret that an increasing number of species may become extinct because of the degradation of the environment by humanity, these data can be best understood in terms of mindless natural selection. The large number of species results from the many, largely random attempts that evolution makes to produce a solution to the survival problem; many failures are to be expected as the bulk of these solutions fail. Many successes are marginal, leaving the species open to eventual extinction. We also now know that mass extinctions have occurred several times as the result of natural catastrophes, such as meteorite strikes or geologic disruptions.

The other place where evidence for the absence of beneficent design can be found is in the short, brutal existences of most life-forms. A common misunderstanding holds that Darwin's discovery of evolution led to his loss of faith. Actually, it wasn't theoretical musings but his lifetime of careful observations of nature. On May 22, 1860, Darwin wrote to American botanist Asa Gray (d. 1888): "I cannot see as plainly as others do, and as I should wish to do, evidence of design and beneficence on all sides of us. There seems to me too much misery in the world. I cannot persuade myself that a beneficent and omnipotent God would have designedly created the Ichneumonidae [wasps] with the express intention of their [larva] feeding within the living bodies of Caterpillars, or that a cat should play with mice."[50]

More recently, Dawkins has written, "The universe that we observe has precisely the properties we should expect if there is, at bottom, no design, no purpose, no evil, no good, nothing but pitiless indifference."[51]

Indeed, Earth and life look just as they can be expected to look if there is no designer God.

NOTES

1. William Paley, *Natural Theology or Evidences of the Existence and Attributes of the Deity Collected from the Appearance of Nature* (London: Halliwell, 1802).

2. Keith Thomson, *Before Darwin: Reconciling God and Nature* (New Haven and London: Yale University Press, 2005), p. 20.

3. Ibid., p. 6.

4. Charles Darwin, *The Origin of Species by Means of Natural Selection* (London: John Murray, 1859).

5. Michael Shermer, *In Darwin's Shadow: The Life and Science of Alfred Russel Wallace* (Oxford, New York: Oxford University Press, 2002).

6. Phillip E. Johnson, *Evolution as Dogma: The Establishment of Naturalism* (Dallas, TX: Haughton Publishing Co., 1990); *Darwin on Trial* (Downers Grove, IL: InterVarsity Press, 1991); *Reason in the Balance: The Case Against Naturalism in Science, Law, and Education* (Downers Grove, IL: InterVarsity Press, 1995); *Defeating Darwinism by Opening Minds* (Downers Grove, IL: InterVarsity Press, 1997); *The Wedge of Truth: Splitting the Foundations of Naturalism* (Downers Grove, IL: InterVarsity Press, 2001).

7. Ronald Numbers, *The Creationists: The Evolution of Scientific Creationism* (New York: Alfred A. Knopf, 1992).

8. John C. Whitcomb Jr. and Henry M. Morris, *The Genesis Flood: The Biblical Record and Its Scientific Implications* (Philadelphia: Presbyterian and Reformed Publishing Co., 1961).

9. William R. Overton, *McLean v. Arkansas*, U.S. Dist. Ct. Opinion, 1982; Michael Ruse, ed., *But Is It Science? The Philosophical Questions in the Creation/Evolution Controversy* (Amherst, NY: Prometheus Books, 1996), pp. 307–31.

10. Michael J. Behe, *Darwin's Black Box: The Biochemical Challenge to Evolution* (New York: Free Press, 1996); William A. Dembski, *The Design Inference* (Cambridge: Cambridge University Press, 1998); *Intelligent Design: The Bridge between Science and Theology* (Downers Grove, IL: InterVarsity Press, 1999); *No Free Lunch: Why Specified Complexity Cannot Be Purchased without Intelligence* (Lanham, MD: Rowman & Littlefield, 2002).

11. Barbara Forrest and Paul R. Gross, *Creationism's Trojan Horse: The Wedge of Intelligent Design* (Oxford and New York: Oxford University Press, 2004).

12. Robert Dorit, review of *Darwin's Black Box* by Michael Behe, *American Scientist* (September–October 1997); H. Allen Orr, "Darwin v. Intelligent Design (Again): The Latest Attack on Evolution Is Cleverly Argued, Biologically Informed—And Wrong," *Boston Review* (1998); Brandon Fitelson, Christopher Stephens, and Elliott Sober, "How Not to Detect Design—Critical Notice: William A. Dembski, "The Design Inference," *Philosophy of Science* 66, no. 3 (1999): 472–88; Kenneth R. Miller, *Finding Darwin's God: A Scientist's Search for a Common Ground between God and Evolution* (New York: HarperCollins, 1999); Robert T. Pennock, *Tower of Babel: The Evidence Against the New Creationism* (Cambridge, MA: MIT Press, 1999); Niall Shanks and Karl H. Joplin, "Redundant Complexity: A Critical Analysis of Intelligent Design in Biochemistry," *Philosophy of Science* 66 (1999): 268–98; Taner Edis, "Darwin in Mind: 'Intelligent Design' Meets Artificial Intelligence," *Skeptical Inquirer* 25, no. 2 (2001): 35–39; James Rachels and David Roche, "A Bit Confused: Creationism and Information Theory," *Skeptical Inquirer* 25, no. 2 (2001): 40–42; Jeffery Shallit, review of *No Free Lunch* by William Dembski, *Biosystems* 66, nos. 1–2 (2002): 93–99; Mark Perakh, *Unintelligent Design* (Amherst, NY: Prometheus Books, 2003); Forrest and

Gross, *Creationism's Trojan Horse*; Matt Young and Taner Edis, eds., *Why Intelligent Design Fails: A Scientific Critique of the New Creationism* (New Brunswick, NJ: Rutgers University Press, 2004). For my own discussion, see Victor J. Stenger, *Has Science Found God? The Latest Results in the Search for Purpose in the Universe* (Amherst, NY: Prometheus Books, 2003), chap. 4. Young and Edis contains a complete listing of current Internet sites discussing both sides of the issue.

13. A review by one of the main promoters of intelligent design has been published in the journal of a small biological society. See Stephen C. Meyer, "The Origin of Biological Information and the Higher Taxonomic Categories," *Proceedings of the Biological Society of Washington* 117, no. 2 (2004): 213–39. The society has publicly repudiated this publication in a statement on September 7, 2004. See http://epsc.wustl.edu/~spozgay/home/id_statement.pdf (accessed July 11, 2006).

14. Behe, *Darwin's Black Box*.

15. Dorit, review of *Darwin's Black Box*; Miller, *Finding Darwin's God*; Perakh, *Unintelligent Design*; David Ussery, "Darwin's Transparent Box: The Biochemical Evidence for Evolution," in Young and Edis, *Why Intelligent Design Fails*, chap. 4.

16. Stephen J. Gould, *The Panda's Thumb* (New York: Norton, 1980), pp. 19–34.

17. H. J. Muller, "Reversibility in Evolution Considered from the Standpoint of Genetics," *Biological Reviews* 14 (1939): 261–80. Another bit of misinformation often bandied about by creationists is that no evolutionary biologist has ever won the Nobel Prize.

18. Richard Dawkins, *The Blind Watchmaker: Why the Evidence of Evolution Reveals a Universe without Design* (London, New York: Norton, 1987), p. 93.

19. Kenneth R. Miller, "Life's Grand Design," *Technology Review* 97, no. 2 (1994): 24–32.

20. Richard Dawkins, *Climbing Mount Improbable* (New York, London: Norton, 1996). See the chapter "The Fortyfold Path to Enlightenment."

21. R. D. Fernald, "Evolution of Eyes," *Current Opinions in Neurobiology* 10, no. 4 (2000): 444–50.

22. Dembski, *The Design Inference, Intelligent Design*, "The Design Inference."

23. For the most recent work at this writing, see the chapters by Gishlack, Shanks, and Karsai; Hurd, Shallit, and Elsberry; and Perakh in Young and Edis, *Why Intelligent Design Fails*.

24. Dembski, *Intelligent Design*, p. 168.

25. Stenger, *Has Science Found God?* pp. 102–10. The connection between information and entropy was shown in C. E. Shannon, "A Mathematical Theory of Communication," *Bell System Technical Journal* 27 (July 1948): 379–423; (October 1948): 623–25. See also Claude Shannon and Warren Weaver, *The Mathematical Theory of Communication* (Urbana: University of Illinois Press, 1949).

26. "Department Position on Evolution and Intelligent Design," Department of Biological Sciences, Lehigh University, http://www .lehigh.edu/~inbios/news/evolution.htm (accessed July 11, 2006).

27. Stenger, *Has Science Found God?* pp. 100–102.

28. Laurie Goodstein, "Intelligent Design Might Be Meeting Its Maker," Ideas and Trends, *New York Times*, December 4, 2005.

29. Matthew J. Brauer, Barbara Forrest, and Steven G. Gey, "Is It Science Yet?: Intelligent Design Creationism and the Constitution," *Washington University Law Quarterly* 83, no. 1 (2005), http://law.wustl.edu/ WULQ/83-1/p%201%20Brauer%20Forrest%20Gey%20book%20 pages.pdf (accessed December 28, 2005).

30. *Kitzmuller, et al. v. Dover Area School District et al.*, Case No. 04cv2688, Judge John E. Jones III presiding, December 20, 2005.

31. Overton, *McLean v. Arkansas*, 1982.

32. Larry Laudan, "Science at the Bar—Causes for Concern," *Science, Technology, & Human Values* 7, no. 41 (1982): 16–19. Reprinted in Ruse, *But Is It Science?*, pp. 351–55.

33. Discovery Institute, http://www.discovery.org/scripts/viewDB/ filesDB-download.php?command=download&id=443 (accessed October 28, 2005).

34. Philip Ball, *The Self-Made Tapestry: Pattern Formation in Nature* (New York, Oxford: Oxford University Press, 2001).

35. John Gribbon, *Deep Simplicity: Bringing Order to Chaos and Complexity* (New York: Random House, 2004).

36. Dembski, *The Design Inference, Intelligent Design*, "The Design Inference."

37. Dembski, *Intelligent Design*, pp. 128–31.

38. Ball, *The Self-Made Tapestry*, pp. 105–107.

39. S. Douady and Y. Couder, "Phyllotaxis as a Physical Self-Organized Growth Process," *Physical Review Letters* 68 (1992): 2098.

40. Stuart Kauffman, *At Home in the Universe: The Search for the Laws of Self-Organization and Complexity* (New York and Oxford: Oxford University Press, 1995).

41. Christoph Adami, *Introduction to Artificial Life* (New York: Springer, 1998).

42. Christoph Adami, Charles Ofria, and Travis C. Collier, "Evolution of Biological Complexity," *Proceedings of the National Academy of Sciences USA* 97 (2000): 4463–68.

43. Martin Gardner, "On Cellular Automata, Self-Reproduction, the Garden of Eden, and the Game of 'Life,'" *Scientific American* 224, no. 2 (1971): 112–17; William Poundstone, *The Rescursive Universe* (New York: Morrow, 1985).

44. Stephen Wolfram, *A New Kind of Science* (Champagne, IL: Wolfram Media, 2002).

45. James Gleick, *Chaos: The Making of a New Science* (New York: Viking, 1987).

46. Nicholas Everitt, *The Non-Existence of God* (London, New York: Routledge, 2004), p. 85.

47. A good example is Dembski, *The Design Inference, Intelligent Design*, "The Design Inference." On the other side of the argument, Dawkins, *The Blind Watchmaker* is also somewhat inconsistent in his use of the term "design."

48. S. Jay Olshansky, Bruce Carnes, and Robert N. Butler, "If Humans Were Built to Last," *Scientific American* (March 2001).

49. Dawkins, *The Blind Watchmaker*.

50. Charles Darwin, *The Correspondence of Charles Darwin* 8, 1860 (Cambridge: Cambridge University Press, 1993), p. 224.

51. Richard Dawkins, *River out of Eden* (New York: HarperCollins, 1995); "God's Utility Function," *Scientific American* (November 1995): 85.

Chapter 3

SEARCHING FOR A
WORLD BEYOND MATTER

For the living know that they shall die: but the dead know not any thing, neither have they any more a reward; for the memory of them is forgotten. Also their love, and their hatred, and their envy, is now perished; neither have they any more a portion for ever in any thing that is done under the sun.

—Ecclesiastes 9: 5–6 (King James Version)

MIND AND SOUL

Almost from the moment that modern humans appeared on the scene tens of thousands of years ago, they seem to have possessed a vague notion that they were more than the physical bodies that were born of women, grew and aged, eventually ceased to move and breathe, and finally disintegrated into a small

77

pile of dusty bones. At some point in their development, people in almost every culture have imagined invisible spirits acting as agents for events around them, including the animation of living things such as themselves.

Such thinking was perfectly reasonable during the childhood of humanity. One moment a person is talking and walking around and in another moment he is forever silent and immobile. Whatever animated the person was suddenly absent. Furthermore, a dead person still seemed to live on in thoughts and dreams—a ghostly spirit surviving death.

A widespread ancient belief held that the heart is the center of being and intelligence. This idea carries metaphorically down to today, as when we say someone has a "good heart" or talk about some act "coming from the heart." When Egyptian priests prepared the dead for their afterlife, they disposed of the brain but kept the heart within the body. Early Greek philosophers, such as Empedocles (d. 490 BCE), attributed thinking and feeling to an immortal soul that resides around the heart but leaves the body after death.

The brain was not regarded as an important organ in ancient times, although Alcmaeon (c. 500) declared, "All senses are connected to the brain." Still, like other ancient Greeks, he viewed the body as containing channels for spirits (*pneumata*) that were composed of air—one of the four elements of the cosmos that included fire, earth, and water. Plato (c. 347 BCE) placed a "vegetative soul" in the gut, a "vital soul" in the heart, and an immortal soul in the head. His most famous student, Aristotle (d. 322 BCE), restored the immortal soul to the heart. Whatever its location, in the common view the soul was a conduit for spirits—the force that gave a body life and thought.[1]

The association of spirit with air is embedded in a number of ancient languages: the Hebrew *ruah* ("wind" or "breath") and *nefesh*, also associated with breathing; the Greek *psychein* ("to breathe"), which is related to the word *psyche* for "soul"; and the Latin words *anima* ("air," "breath," or "life") and *spiritus*, which also refers to breathing.[2] The soul was seen as departing the body in the dying last breath.

In Hawaii, native shamans attempted to breathe life back into a dead body by shouting "ha!" Western doctors were seen not to do this and so were said to be "ha-ole"—without ha. In today's diverse population in Hawaii, Caucasians are commonly called *haoles*.

In the Old Testament, the soul is life itself, breathed into the body by God. While traditional Judaism does not regard death as the end of human existence, it has no dogma of an afterlife, and a range of opinions can be found among Jewish scholars. Christianity, on the other hand, made human immortality its foundational principle, the doctrine probably most responsible for the long success of that faith. The power of Islam can also be attributed to the promise of an afterlife, with dark-eyed maidens providing eternal pleasure (for men, anyway).

Following the teaching of the Greek physician Galen (d. 201), early church fathers located the immortal soul in the empty spaces of the head. However, Christendom lost touch with Greek philosophy after the fall of Rome in 476 until the ancient writings were recovered in the twelfth century, mostly from Islamic sources.[3]

Christians did not take well to the teachings of the Greek atomists, who challenged the whole notion of an immortal soul. Epicurus (d. 270 BCE) taught that the soul was made of matter, like everything else. The soul atoms were concentrated in the chest and took life with them when a body died. In *De Rerum Natura* (*On the Nature of Things*), the Roman poet Lucretius (d. 55 BCE) wrote, "Death is therefore nothing to us and does not concern us at all, since it appears that the substance of the soul is perishable. When the separation of body and soul, whose union is the essence of our being, is consummated, it is clear that absolutely nothing will be able to reach us and awaken our sensibility, not even if earth mixes with sea and seas with heaven."[4]

Most laypeople today take for granted a separateness or "duality" of soul and body, of spirit and matter. However, this distinction was not made clear-cut until the seventeenth century, when René Descartes (d. 1650) found a way to reconcile atoms and soul. This was the age when machines were coming into common use. Descartes was a contemporary of Galileo Galilei (d. 1642), two generations ahead of Isaac Newton (d. 1727). The French thinker developed many of the mathematical methods such as representing curves by equations and the Cartesian coordinate system that would receive wide application in the new science of mechanics that was elaborated by Newton.

Descartes argued that animals, including humans, were intricate, material machines—designed by God, of course (he was terrified of the Inquisition). However, he argued that humans possess an additional ingredient that is not composed of the basic particles of matter: an immaterial soul. The soul did everything that machines were presumably not capable of doing: thinking, consciousness, will, abstraction, doubt, and understanding.[5] Descartes speculated that the pineal gland of the brain marked the place where the soul and the brain interacted.

Descartes was also a contemporary of Thomas Hobbes (d. 1679), who agreed with him on the machinelike nature of the human body but viewed the notion of an additional, immaterial soul as a delusion. Hobbes even went further in proposing that society itself could be understood as a clockwork mechanism and, in his most famous work, *Leviathan*, first published in 1652, he attempted to deduce the optimum political structure. He determined it to be dictatorship, by a king or otherwise.[6]

At this significant turning point in history, empirical science in Europe was beginning to raise doubts about the blind obedience to authority that had stifled progress for centuries. Copernicus and Galileo had based their new cosmology, which challenged the teachings of Aristotle, on empirical data—setting the stage for the Newtonian revolution. But, even before that happened, a brave new breed of empiricists was taking a closer look at the bodies of humans and animals.

RISE OF THE BRAIN

In a fascinating book, *Soul Made Flesh: The Discovery of the Brain—and How It Changed the World*, Carl Zimmer tells the story of a remarkable group of seventeenth-century men working in Oxford during the English Civil War and its aftermath, who by dissecting human and animal cadavers established, among numerous other anatomical facts, that the brain was the primary organ of thought.[7] These included several who became famous for other individual achievements: Christopher Wren (d. 1723) designed the magnificent Sheldonian Theater in Oxford while drafting detailed illustrations of human organs. Robert Boyle (d. 1691) transformed alchemy to modern chemistry and demonstrated the pressure of air, while conducting hundreds of experiments on anatomy. Boyle's assistant Robert Hooke (d. 1703) discovered the law of springs while designing instruments such as a microscope that enabled investigators to see the intricate structures inside living organisms.

The leader of the "Oxford Circle" was a physician, Thomas Willis (d. 1675), who produced the first detailed anatomy of the brain and traced the nervous system throughout the body. He identified the heart as a blood pump that operated under the control of signals from the brain. Like his contemporaries, Willis referred to these signals as "spirits." Not until the eighteenth century would the signals carried by nerves be identified with electricity.

After the restoration of Charles II to the throne, the Oxford Circle came out into the open, moved to London, and evolved into the Royal Society for Promoting Natural Knowledge, which became a catalyst for the scientific revolution that followed.

Willis founded the science of neurology, which eventually confirmed many of his notions, at least in a general way. We now know that electrical impulses compose the "spirits" that carry signals from the brain through the nervous system. Different parts of the brain perform different functions. The human brain is basically similar to that of other animals, differing in those portions that give us our superior cognitive and intellectual abilities. Psychological disorders arise in the brain and are routinely treated today with chemicals. And, as we all are well aware, chemicals can also cause mental disorders or alter mental states and even trigger "spiritual experiences" (as with LSD). Brain diseases, such as Alzheimer's, affect memory and behavior. All of this strongly implies that our thoughts, memories, and subjective experiences may be entirely based upon physical processes in the brain.

BRAIN SCIENCE TODAY

Scientists no longer need to remove the brain from a dead body in order to study it. Imaging technology makes it possible not only to examine brains in detail but also to observe them while they are still alive and functioning. In recent years, this has enabled the sources of perceptual judgments and different types of thought to be located within the brain. Experiments have been conducted in which subjects are asked to make mechanical, intellectual, and moral choices, while researchers watch the brain carry out the necessary operations.

A number of imaging techniques have been developed with modern technology. Perhaps the most powerful is *magnetic resonance imaging* (MRI). Based on the physics of *nuclear magnetic resonance* (NMR), with the word "nuclear" removed so as not to alarm patients, MRI forms an image by detecting the energy that is released by the spinning nuclei of atoms. This energy is actually very low, coming from the radio region of the electromagnetic spectrum and not at all harmful—especially compared to x-rays, which have sufficient energy to break atomic bonds. In *functional* MRI (fMRI), the magnetic properties of the blood are used to see patterns of blood flow. An fMRI scan of the brain can quickly produce images that distinguish structures less than a millimeter apart and pinpoint areas in the brain that are being activated.

Other brain imaging techniques include *positron emission tomography* (PET), *single photon emission computed tomography* (SPECT), and *electroencephalography* (EEG).[8]

All these techniques confirm that thought processes are accompanied by localized physical activity in the brain. Let us look at just a few of the examples relevant to our discussion. Many more can be found in the literature.

Using fMRI, scientists in the United States and Brazil have discovered that the region of the brain activated when moral judgments are being made is different from the region activated for social judgments that are equally emotionally charged.[9] Princeton researchers have studied the brain activity in people asked to make decisions based on various moral dilemmas. These dilemmas were divided into two categories—one involving impersonal actions and another where a direct personal action was required. The brain scans consistently showed greater activation in the areas of the brain associated with emotions when the actions were personal.[10] The relevant point here is not just that physical processes in the brain take part in thinking; they seem to be responsible for the deepest thoughts that are supposed to be the province of spirit rather than matter.

Another area of study with live brains involves the localized stimulation by electric or magnetic pulses. Neuroscientist Michael Persinger claims to have induced many of the types of experiences that people have interpreted as "religious" or "spiritual" by magnetic stimulation of the brain.[11] However, Persinger's results have been called into question.[12]

On the other hand, Olaf Blanke and his colleagues report that they are able to bring about so-called *out-of-body experiences* (OBE), where a person's consciousness seems to become detached from the body, by electrical stimulation of a specific region in the brain.[13] I have discussed OBE experiments in two books and have concluded that they provide no evidence for anything happening outside of the physical processes of the brain.[14]

These results do not totally deny the possibility that conscious thoughts are being directed by a disembodied soul, which then somehow implements them through the brain and nervous system. This, in one form or another, remains the teaching of most religions. In 1986 Pope John Paul II reaffirmed the 1950 statement by Pope Pius XII that the Church does not forbid the study and teaching of biological evolution.[15] However, the pope made it very clear that evolution applied to the body—not the mind: "Theories of evolution which, in accordance with the philosophies inspiring them, consider the spirit as emerging from the forces of living matter, or as a mere epiphenomenon of this matter, are incompatible with the truth about man. Nor are they able to ground the dignity of the person."[16]

Despite the Holy Father's admonition, a wealth of empirical data now strongly suggests that mind is in fact a "mere epiphenomenon of this matter." Matter alone appears to be able to carry out all the activities that have been traditionally associated with the soul. No "spiritual" element is required by the data. The implication that "we" are bodies and brains made of atoms and nothing more is perhaps simply too new, too disturbing, too incompatible with common preconceptions to be soon accepted into common knowledge. However, if we do indeed possess an immaterial soul, or a material one with special properties that cannot be found in inanimate matter, then we should expect to find some evidence for it.

Hundreds of reports of scientific observations of special powers of the human mind under claimed "controlled conditions" have been made over the past one hundred and fifty years. Not a single one has met all five of the conditions, listed in chapter 1, that are required for science to take an extraordinary claim seriously. Are these conditions unreasonable? Am I asking too much of the investigators? I can list dozens of extraordinary scientific discoveries made during that same period that have met precisely these same conditions, so this cannot be attributed to some dogmatic bias in science against "new ideas."

Obviously I cannot do a survey of every claim, although in my 1990 book, *Physics and Psychics*, I singled out for critical analysis those that the proponents themselves considered the most convincing.[17] These were brought up to date in my 2003 book, *Has Science Found God?*[18] In what follows I will review some sample claims that should sufficiently illustrate why the case for special powers of the mind has not been made.

THE FORCE OF LIFE

Let us begin by considering the ancient association of the soul with life itself, as a kind of special ingredient, an *élan vital* or vital force, that is possessed by live organisms and was long thought to distinguish them from inanimate objects such as rocks and dead organisms. Many cultures have held such beliefs, and even today we hear terms like *qi* (chi) used to represent some special energy that is supposed to flow through the body. In Western religions this life force is often identified with the soul. If such a life force exists, then we should be able to detect its presence.

Although much of complementary and alternative medicine (that is, nonscientific therapies) is based on the assumption of a life force, sometimes called the "bioenergetic field," biological science has not uncovered its presence within humans, animals, or plants.[19] Well-understood physical and chemical processes, the same that occur in all materials whether dead or alive, are sufficient to account for the observed interactions between various parts of living organisms. The physics and chemistry of living cells is basically the same as the physics and chemistry of rocks, just a bit more complicated.

The sensitive detectors used in physics laboratories are

capable of detecting various kinds of radiation of very low intensity. Except for some weak electromagnetic radiation emitted by oscillating charges in the heart and brain that can be picked up with sensors placed directly on the skin, and the infrared thermal radiation emitted by all physical bodies dead or alive (or never alive, like rocks), living organisms emit no unique radiation that can be detected by our best scientific instruments.

Of course, one can argue that the instruments are simply insensitive to "living energies," although the proponents of bioenergetic fields generally claim a connection to easily detectable electromagnetic waves.[20] If it is significant, some effect should be measurable. For example, a widely used therapy is called *therapeutic touch*, in which a healer "manipulates" a patient's "energy field." After a decade or so of common use, you would reasonably expect some evidence for the efficacy of the treatment. In fact, there is none that is not purely anecdotal and thus not amenable to proper scientific testing. Indeed, therapeutic touch has been tested and failed the test.[21]

QI = MC²?

I recently examined a published claim that the vital force called qi has been demonstrated in a scientific experiment in China. I presented my analysis at several universities in China during a visit there in April 2005 as part of a scientific delegation.

The reported experiments were performed during several public healing "lectures" by qi master and healer Dr. Xin Yan in Beijing in 1987 and published in a peer-reviewed American journal.[22] Positive signals above background levels, claimed to be qi, were reported in standard radiation dosimeters. Both the background levels and signals were quite high. Other phenomena were reported that I did not consider because the paper lacked sufficient information.

The Yan paper was not published until 2002 and makes no mention of any successful (or unsuccessful) attempts at replication during the intervening years. The results are difficult to evaluate from the data presented. Furthermore, no error estimates are given, which would be sufficient cause to deny publication in most reputable scientific journals.

Nevertheless, the data presented are sufficient in the case of one experiment to draw some conclusions. In this experiment, Dr. Yan "emitted qi" during an eleven-hour (!) "lecture." *Thermoluminescence dosimeters* (TLDs) of the type commonly used in nuclear laboratories to measure radiation exposure were placed throughout the auditorium. Doses significantly above background were reported from different directions, indicating that the supposed qi-rays were unfocused. Although some of the other experiments contained controls, no measurements taken under identical conditions with the qi master absent are reported for this particular experiment.

In figure 3.1, I have plotted the reported dosages measured by two types of TLDs as a function of distance from the podium. One type [^{7}LiF(Mg,Ti)] is sensitive to gamma rays while the other type [^{6}LiF(Mg, Ti)] is sensitive to thermal neutrons as well as gamma rays. I averaged over the two sides of the auditorium where the intensities were comparable. The squares and circles on the figure show the measured radiation exposure in milliroentgens (mR) accumulated over the eleven-hour experiment. For gamma rays, one milliroentgen is approximately equivalent to one millirem (mrem), the unit used to measure biologically significant exposure. If the numbers are accurate, they represent an intensity that would exceed the generally considered safe dosage if experienced steadily for a year, five thousand mrem. That is, the recorded radiation intensity was appreciable.

At the same time, the dosimeters used in the experiment are designed for measuring long-term accumulated exposure with about a ten-mR detection limit. They were not particularly suit-

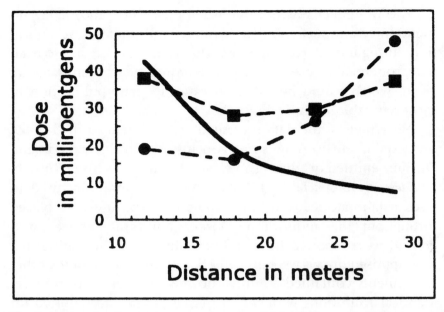

Fig. 3.1. Results from the experiment of Xin Yan et al. The square points are the data from the dosimeters sensitive to neutrons and γ-rays. The round points are γ-rays only. The solid curve shows what would be expected if the measured radiation were conserved as would be expected for any form of energy.

able for the short-term exposures used here, and more precise instruments for measuring instantaneous radiation intensities are readily available. As mentioned, no estimates or errors are given in the paper (sufficient cause for its rejection). If we put ten mR error bars on the data points, the results are insignificant.

The authors claim numerous reports from the audience of beneficial health effects, although they present no data on this. Gamma rays and neutrons are not noted for their positive health consequences unless directed at tumors, and the authors concede, "It is highly unlikely that the qi field generated by Dr. Yan contains actual gamma rays and neutrons. Rather the TLD readings seem to be a phenomenological description of the interactions between a TLD detector and Dr. Yan's qi field." They offer no theoretical model for the phenomenon, no suggestion on how qi-rays might affect these particular detectors.

Independent of the significance of the dosage level, we see in figure 3.1 that the "gamma-ray" data actually increase with distance, while the "neutron-plus-gamma" data show no significant distance effect. The smooth curve plotted on the same graph shows the (unobserved) falloff with the square of distance that is required by energy conservation (arbitrary scale).

If you were to ask me, "What is the defining property of energy?" I would answer the fact that it is conserved. If energy were not conserved, the quantity would be of little use in physics. When one measures a quantity that is not conserved under conditions when it should be, then that can be taken as good evidence that what is being observed is not some form of energy. Qi does not look like energy. Indeed, it looks nonexistent.

ESP

One special ability of minds that is widely taken as real (especially in science fiction) would be *extrasensory perception* (ESP), in which minds communicate with one another by some mechanism that is not at present part of established scientific knowledge. Another is *psychokinesis* (PK), or mind over matter, where thoughts are capable of moving objects or otherwise affecting physical phenomena—in the past, present, and future. If a disembodied soul can use some from of psychokinesis to move around brain molecules, then it should be equally well able to move around molecules outside the brain.

If these phenomena exist, then they should be readily detectable in controlled, scientific experiments. Since the mid-nineteenth century scientists have attempted to scientifically verify the reality of unusual mental phenomena. These included the prominent physicists Michael Faraday, William Crookes, and Oliver Lodge. Faraday, the greatest experimentalist of the day, found no evidence, while Crookes and Lodge convinced themselves that they had discovered what they called the *psychic force.*

However, Crookes and Lodge did not control their experiments sufficiently to make them convincing.[23] They generally worked with spiritual "mediums" who were highly skilled at the various illusions that professional magicians and charlatans have developed over the centuries.

Crookes, Lodge, and other early psychic investigators made a fundamental error in allowing their subjects to control the protocol of their experiments. Even today we find this serious breach of commonsense methodology routinely made in psychic experiments. For example, consider the much-touted experiments conducted at the Princeton Engineering Anomalies Research Laboratory (PEAR).[24] Scientists are no more capable of uncovering trickery than anyone else not specifically initiated into the magical arts—perhaps even less so since they are not used to the universe lying to them. Crookes and Lodge proved to be particularly gullible, possibly because of personal tragedies in their lives.[25]

The need for better controls in psychic experiments was recognized in the 1930s by botanist Joseph Banks Rhine of Duke University. Rhine coined the term *ESP* and made an honest attempt to find empirical evidence for the existence of psychic forces. He announced a number of claims that did not stand up to critical scrutiny and, after numerous rejections by established scientific journals, he started his own journal for which he could choose more sympathetic reviewers. Despite his failure to convince mainstream scientists of the reality of psychic forces, Rhine pioneered a field of study that continues to the present day under the designation of *parapsychology.*[26] Even parapsychologists must admit that they operate on the borders of conventional science.

As I have mentioned, there is no agreed-upon precise definition of science. So I will not press the point as to whether or not parapsychology is science. Parapsychologists continue to make claims that ESP has been observed in controlled experiments. Some of these reports are peer reviewed, but the peers are generally other true believers who review manuscripts for special journals like Rhine's that maintain different standards than mainstream scientific journals. The editors of these journals claim they provide a greater "openness" to new ideas. This is fine, but the publishing of poorly executed experiments, as exemplified by the qi experiment described above, does not serve any useful purpose and drags down the credibility of everything else published by that journal.

As with the creationists described in chapter 2, proponents of ESP claim that their results are unfairly rejected because of conventional science's dogmatic attachment to old ideas. My reaction is the same as it was in the case of intelligent design: what possible reason would scientists have to object if convincing evidence for psychic phenomena was reported? As with intelligent design, the discovery of special powers of the mind would open up wonderful new avenues of research that would surely be generously funded by taxpayers. Mainstream scientists have not accepted the claims of parapsychology for exactly the same reason they have not accepted the claims of intelligent design. The data do not warrant it.

From the first experiments in the mid-nineteenth century to the present, the claim of evidence for ESP simply does not stand up under the same scrutiny scientists apply when considering any extraordinary claim.

THE SIGNIFICANCE OF EXPERIMENTS

Let me expand on the issue of statistical significance of experiments, which is the basis on which many reported extraordinary claims can be quickly discarded. Parapsychologists argue that

they should be held to the same standard of statistical significance as medical science, where claimed positive effects of, say, a
new drug, are published when the statistical significance ("P
value") is 5 percent ($P = 0.05$) or lower. That is, if the experiment
were repeated many times in exactly the same fashion, on average
one in twenty would produce the same effect, or a greater one, as
an artifact of the normal statistical fluctuations that occur in any
measurement dealing with finite data.

But think of what that means. In every twenty claims that are
reported in medical journals, on the average one such report is
false—a statistical artifact!

Contrast this with the standard in the field of research where I
spent my career, elementary particle physics. There the standard of
P value for publication of an important new discovery is one-
hundredth of one percent ($P < 0.0001$). This guarantees that, on
average, only one in ten thousand such reports is a statistical artifact.

A possible justification for the low standard in medicine may
be that medical journals are not venues for extraordinary new
discoveries but places where promising new therapies are dissem-
inated to the healthcare community as rapidly as possible. If one
in twenty are spurious, that may be regarded by some as a small
price to pay if a life might be saved by a therapy that works. Nevertheless, I think the medical standard should be higher, given
the large number of false reports that are later withdrawn. Think
of all the wasted money, effort, and lives that probably go into
useless therapies under the current arrangement.

Indeed, medical researchers are beginning to recognize the
inadequacy of their journal standards. Epidemiologist John Ionnidas has gone so far as to write, "Most published research findings in Medicine are false."[27] A recent paper in *British Medical
Journal* recommends the P value threshold be changed to $P <
0.001$, not as tight as in physics but probably suitable for medical
science, given all its added complications.[28]

Parapsychologists, on the other hand, are not in the business of saving lives. They are more like particle physicists or astronomers, seeking to uncover facts about the fundamental structure of nature, where no one will die if the report of an important discovery is held off for a few months or years.

Almost without exception, claims of evidence for psychic phenomena come nowhere close to having the statistical errors small enough to rule out more mundane explanations for the results.[29] The handful that claim reasonable statistical significance all have methodological flaws that render their results unconvincing. And none are independently replicated at a statistically significant level.

A number of studies have claimed to be able to overcome the lack of statistical significance of single experiments by using a technique called "metanalysis," in which the results of many experiments are combined.[30] This procedure is highly questionable.[31] I am unaware of any extraordinary discovery in all of science that was made using metanalysis. If several, independent experiments do not find significant evidence for a phenomenon, we surely cannot expect a purely mathematical manipulation of the combined data to suddenly produce a major discovery.

No doubt parapsychologists and their supporters will dispute my conclusions. But they cannot deny the fact that after one hundred and fifty years of attempting to verify a phenomenon, they have failed to provide any evidence that the phenomenon exists that has caught the attention of the bulk of the scientific community. We safely conclude that, after all this effort, the phenomenon very likely does not exist. In any other field, such an unbroken history of negative results would have long ago resulted in the claims being discarded. At the minimum, psychic experiments cannot be used to show that humans possess any special powers of the mind that exceed the physical limitations of inanimate matter.

DOES PRAYER WORK?

One of the defining characteristics of the Judeo-Christian-Islamic God is that he is believed to respond to entreaties from the faithful and steps in to change the natural course of events when he is sufficiently moved by the intensity and piety of the petitioner (or, whenever he wishes). Surely, with the millions of prayers being submitted daily, totaling billions in recorded history, some objectively verifiable (not just anecdotal) positive evidence should have been found by now!

Of course, prayer by or in the presence of a patient plausibly could have some purely natural beneficial effects, such as helping relax an ill person, lower blood pressure, and so on. However, this effect is small at best and indistinguishable from other forms of relaxation that contain no religious or spiritual element.[32] Actually, as we will see, some data suggest that such prayer may actually be detrimental, possibly adding to the anxiety of the patient. In any case, to be considered extraordinary evidence in favor of prayer, experiments must be "blinded" so that neither patients nor investigators know who is being prayed for.

It might seem that prayer is not amenable to scientific testing. First, it is supposedly "spiritual" rather than material. Second, prayer is difficult to control. For example, how could you stop someone from praying or know for sure that a subject is not being prayed for somewhere in the world? However, anything with observable consequences is testable by scientific means, and prayer is widely believed to have observable consequences. A positive signal is possible if, for example, some type of prayer is superior to another. This would show up in a statistically and systematically significant success rate for that type. In chapter 1 I presented a hypothetical example where Catholic prayers were convincingly demonstrated to work in careful scientific experiments, while those of other religions failed. It would be difficult to think of a plausible natural mechanism for this phenomenon.

As already noted, despite official statements from some national science organizations, science is not wholly restricted to the consideration of purely material causes for observable phenomena. If empirical data show some result that cannot be accounted for by current, conventional materialistic means, then good science as well as honesty demands that this fact be acknowledged and published. The issue of whether no material mechanism can ever be found could be left open for further research, which would surely get funded—once again leaving scientists happy as clams.

The effects of prayer should be readily measurable, in particular where prayers may be focused on some specific purpose such as healing the afflicted. As we saw with psychic phenomena, many popular books and articles have been published claiming that science has shown that prayer has positive healing value.[33] But, once again, we find that none of the reports is convincing. I discussed several specific examples in *Has Science Found God?* and will not repeat these here.[34] Every published claim of a positive effect of which I am aware fails to satisfy one or more of the methodological conditions I laid out in chapter 1. As I have emphasized, these conditions are routinely applied to all extraordinary claims in physics or other "hard" sciences. With all the publicity that attends to prayer studies, it is highly unlikely any good quality study has been missed.

Since my previous book went to press, several important new results have been published that have virtually settled the matter. One case, in particular, has generated considerable attention and provides valuable insight into the admitted difficulties that arise when the attempt is made to use rational science to evaluate long and deeply held religious beliefs. But then we will see that when scientists do their jobs properly, not allowing their personal beliefs to overrule their objective analysis of the data, we can have confidence in their results.

THE COLUMBIA "MIRACLE" STUDY

In 2001 the *Journal of Reproductive Medicine* published an article submitted by the highly prestigious Columbia University Medical Center claiming to show that infertile women who were prayed for by Christian prayer groups became pregnant twice as often as those not prayed for.[35] This caught the immediate attention of national media, including ABC News, whose medical editor, Timothy Johnson, credulously reported on the "surprising results" to millions on *Good Morning America*.[36] It is probably not irrelevant to mention that Johnson at the time was also serving as a minister at the evangelical Community Covenant Church in West Peabody, Massachusetts.

The study was actually not conducted at Columbia but rather in Korea at an institute directed by one of the three coauthors, Kwang Cha. A sample of 219 women was separated randomly into two groups, one of which was prayed for and the other not. Christian prayer groups in the United States, Canada, and Australia conducted the prayers, with the investigators masked until the data were all collected and the clinical outcomes known.

The reported results showed that the prayed-for group had a pregnancy rate of 50 percent, while the not-prayed-for group had only 26 percent. The statistical significance for the difference was $P = 0.0013$. The prayed-for group also had a higher success rate for *in vitro fertilization-embryo transfer*, 16.3 percent compared to 8 percent, $P = .0005$.

While the first result does not quite meet the new standard of $P < 0.001$ suggested above, these statistical significances are certainly better than the worthless $P = 0.05$ we usually see. At the very least, if this report is correct, then attempts at replication are reasonably justified.

However, doubt has been cast on the validity of the results. Bruce L. Flamm, clinical professor of obstetrics and gynecology at the University of California, Irvine, found a number of flaws in the study protocol, calling it "convoluted and confusing."[37] For example, one group of prayer participants prayed directly for the patients while a second group not only prayed for the patients but also prayed for the effectiveness of the prayers of the first group. A third group simply prayed that "God's will or desire be fulfilled," whatever that is.

Perhaps these confusions are not too serious and, in any case, could be easily rectified in a follow-up study. However, further convolutions and confusions have been revealed about the participants in the study.

One of the authors, Daniel P. Wirth, is a lawyer without a medical degree. However, he does have a degree in parapsychology and has authored several articles in parapsychology journals claiming documented evidence for faith healing.[38] In an unrelated matter, Wirth has since been imprisoned after being convicted of fraud, which included the use of names of dead people for financial gain.

The lead author of the paper was originally identified as Rogerio Lobo, then head of the Columbia University department of obstetrics and gynecology. However, shortly after publication, Columbia University announced Lobo was not even aware of the study until being informed by Cha six to twelve months after the study was completed. Lobo has since withdrawn his name from the study and any connection between Cha and Columbia has been severed. The paper, however, has not been formally withdrawn—a black mark on a great university.

Neither Columbia University nor the *Journal of Reproductive Medicine* has come completely clean on this fiasco, and while some media outlets have reported on the questionable nature of the claims, that knowledge has not become as widespread as the astounding claims made in the original announcement. What is referred to as the "Columbia miracle study" continues to be referred to by shameless promoters of faith healing, such as Larry Dossey, as one of his exemplars of the "controlled clinical trials and peer-review process" that provide scientific support for the efficacy of prayer.[39] Indeed the experiment was exemplary. It serves as a prime example of how not to conduct a scientific investigation of extraordinary claims.

CAN PRAYER CHANGE THE PAST?

Dossey was also impressed by a study reported in the *British Medical Journal* in 2001 reporting that praying for patients reduced their length of stay in hospital ($P = 0.01$) and duration of infections ($P = 0.04$).[40] If this was not remarkable enough, the prayers were actually performed *after* the patients had left the hospital, implying that the power of prayer extends into the past as well as the future. Note that the journal did not apply the $P < 0.001$ standard that it, itself, had proposed that same year (see discussion above).

It is not clear how seriously the author of this report, Dr. Leonard Leibovici, meant for us to take his results. He had earlier declared, "Empiricists are not equipped to recognize the loud signals of alternative medicine as false," calling alternative (complementary) medicine a "cuckoo in the nest of . . . reed warblers."[41]

Leibovici might regard Larry Dossey and Brian Olshansky as "cuckoo" for taking his report very seriously. They suggest that this result may be reconciled with our present understanding of the universe by going "beyond the superstring theories of today's physicists."[42]

Physician (and devout Christian) Jeffrey P. Bishop and I evaluated these claims in a paper published in 2004 in the *British Medical Journal*, where the other reports had appeared.[43] First, we pointed out that none of the studies in medicine and parapsychology that Olshansky and Dossey take as "confirmatory evidence" are significant. Second, we showed that nothing in modern physics suggests a physical basis for the type of backward causality being suggested.

I have written extensively on the misuse of modern physics, in particular quantum mechanics, to support mystical claims.[44] I have also argued that the results of some physics experiments may be interpreted as evidence for events in the future affecting events in the past.[45] But this only happens at the quantum level, and no theoretical or empirical basis exists for backward causality on the large scale of human experience.

In short, neither robust data nor existing physical, chemical, biological, or neurological theories support the notion that prayer can affect human health—forward or backward in time.

THE DUKE STUDY

Two of the studies I reported on in *Has Science Found God?* involved praying for the improved health of coronary patients.[46] While claiming positive results, neither study produced statistically significant effects, and both experiments were also otherwise severely flawed, so they may be safely discarded. These highly publicized reports have been followed by two well-executed experiments that seem to meet all the requirements of a proper investigation. Both find no evidence that prayer improves health.

In a three-year clinical trial led by Duke University physicians, the effects of intercessory prayer and other so-called noetic therapies such as music, imagery, and touch therapy were examined for 748 patients in 9 hospitals in the United States. Twelve prayer groups from around the world were involved, including lay and monastic Christians, Sufi Muslims, and Buddhist monks. Prayers were even e-mailed to Jerusalem and placed on the Wailing Wall.

Patients awaiting angioplasty for coronary artery obstruction were selected at random by computer and sent to the twelve prayer groups. The groups prayed for complete recovery of the patients. The clinical trials were double blind: neither the hospital staff nor the patients knew who was being prayed for.

The findings, reported in the journal *Lancet*, showed no significant differences in the recovery and health between the two groups.[47] The result for touch therapy was also negative, while the other techniques showed "some promise."

It is notable that this study was not conducted by a bunch of "closed-minded skeptical materialistic atheists" but rather physicians of religious faith who personally believe that alternatives to conventional scientific medicine are worth pursuing. There can be little doubt what, in their hearts, they wanted to see. The lead author, Mitchell Krucoff, was ecstatic when the first results started coming in. In November 2001 he told a media outlet: "We saw impressive reductions in all of the negative outcomes—the bad outcomes that were measured in the study. What we look for routinely in cardiology trials are outcomes such as death, a heart attack, or the lungs filling with water—what we call congestive heart failure—in patients who are treated in the course of these problems. In the group randomly assigned to prayer therapy, there was a 50 percent reduction in all complications and a 100 percent reduction in major complications."[48] But as the significance of the data improved, the situation turned out otherwise. Since he signed the paper, Krucoff is now apparently satisfied with the published conclusion that no effect of prayer has been observed.

A coauthor of the *Lancet* paper was Harold Koenig, who directs the Center for Spirituality, Theology and Health at Duke University in which Krucoff and other coauthors are participants. Koenig is the author of over a dozen book books on healing and faith.[49] There can be no doubt that Koenig, also a person of faith, would like nothing better than to announce the discovery of evidence for the supernatural healing power of prayer. But Koenig is an honest and competent scientist who is not going to make such an announcement until the data warrant it. I have communicated extensively with him and find we have little disagreement on the fact that, after extensive experimentation, any positive benefits of prayer and other religious exercises that may be currently indicated can be understood in terms of physical processes alone. He is also in agreement with Bishop's and my refutation of the claims of efficacy for retroactive prayer.

THE STEP PROJECT

Perhaps the definitive work is the mammoth STEP project (Study of the Therapeutic Effects of Intercessory Prayer), a collaboration of six medical centers, including Harvard and the Mayo Clinic, lead by Harvard professor Herbert Benson.[50] This study, lasting for almost a decade, involved 1,802 patients who were prayed for over a fourteen-day period starting the night before receiving coronary artery bypass graft (CABG) surgery.

The patients were randomly and blindly divided into three groups: 604 received intercessory prayers after being informed they might or might not receive such prayers, 597 did not receive prayers after being informed they might or might not receive such prayers, and 601 received intercessory prayers after being informed they definitely would be prayed for. None of the doctors knew who was being prayed for in the first two groups. Two Catholic groups and one Protestant group carried out the praying. It apparently did not occur to the investigators to also include a group of atheists thinking nice thoughts.

The published results showed that in the two groups uncertain about receiving intercessory prayer, complications occurred in 52 percent (315/604) of patients who received intercessory prayer versus 51 percent (304/597) of those who did not. Complications occurred in 59 percent (352/601) of patients certain of receiving intercessory prayer compared with the 52 percent of those uncertain of receiving intercessory prayer. Major events and thirty-day mortality were similar across the three groups.

The authors concluded that intercessory prayer itself had no effect on complication-free recovery from CABG, but certain knowledge of receiving intercessory prayer was associated with a higher incidence of complications. The later effect somewhat surprised the investigators, who speculated that these patients may have experienced higher anxiety, perhaps thinking they were so desperately ill that they needed to be prayed for. No one suggested that God was deliberately thwarting the expectations of the researchers. Actually, I do not regard this effect as significant.

The investigators included a Catholic priest, Father Dean Marek, who was principal investigator of the Mayo Clinic portion of the study, and other believers. Primary funding of $2.5 million was provided by the John Templeton Foundation, which seeks to find connections between religion and science, so skeptics cannot be blamed for deliberately producing negative results. They were not even involved. Father Marek and other coauthors have tried to account for why prayers do not work within a theological context, but they are to be commended for accepting the data and admitting they did not work in their particular experiment.

As was the case for the special powers of the mind termed "psychic," studies of the supernatural powers of prayer have so far produced no convincing results. If prayer were as important as it is taken to be by Jews, Christians, and Muslims, its positive effects should be obvious and measurable. They are not. It does not appear—based on the scientific evidence—that a God exists who answers prayers in any significant, observable way.

IMMORTALITY

For many if not most believers, the greatest appeal of religion is the promise of eternal life. St. Paul said, "And if Christ not be risen then is our preaching in vain, and your faith is also vain."[51]

In his classic work *The Illusion of Immortality*, philosopher Corliss Lamont surveyed all the aspects of the subject of immortality, from theological and philosophical to scientific and social.[52]

He points out that the exact nature of the immortality that is preached in Christianity, as well as in other religions, is not at all clear, with many different doctrines being presented over the ages.

Part of the problem is one that we can recognize from the earlier discussion on the brain. What is it exactly that survives death? We have seen that neurological and medical evidence strongly indicates that our memories, emotions, thoughts, and indeed our very personalities reside in the physical particles of the brain or, more precisely, in the ways those particles interact. So this would seem to say that when our brains die, we die.

Historically, the Catholic Church has taught that the full body is resurrected. The Apostles' Creed, adopted in the second century and still recited, states that there will be a resurrection of the flesh. The Council of Trent in the sixteenth century asserted that the "identical body" will be restored "without deformities." St. Augustine declared that "the substance of our bodies, however disintegrated, shall be entirely reunited."[53]

This doctrine would seem to satisfy any objection raised by recognition of the physical nature of mind. God simply reassembles us—brain and all—and the brain contains our personalities. Presumably, in heaven we will look as we did at eighteen, but we can hardly expect the same brain that was in our bodies at that age. Heaven forbid! I guess we get the brain we die with, so we have all our memories. But, then, what if we die with Alzheimer's disease?

We need not go any further into these unconfirmable speculations (at least unconfirmable in this life). The scientific question is whether there is any evidence for life after death. As with ESP and other proposed super powers of the mind, despite numerous claims over the years, no claimed connection with a hereafter has ever been scientifically verified. And, as with those special powers, we can easily see how a connection should have been verified in controlled, scientific experiments.

Consider the case of psychics or mediums who claim they have the power to speak to the dead. Such spirits surely would have access to a deep store of information from which some observable phenomenon currently unknown to science can be extracted that could not have been in the psychic's head all along.

For example, suppose a psychic informs his client that her dead mother told him where to find a long-lost engagement ring—behind the kitchen stove. If the ring is then found at that place, it would indeed seem to be miraculous.

However, before accepting this result as confirmation of the extraordinary hypotheses of life after death and the psychic's power to communicate with the dead, you have to rule out all possible ordinary explanations. For example, the psychic may have visited his client at home at some earlier time, seen the ring sitting alongside the sink where it had been removed to wash dishes, and surreptitiously dropped it behind the stove (yes, psychics have been known to cheat). That, and similar possibilities, would have to be ruled out first. But, if properly designed, experiments proving immortality are in principle possible. All that has to happen is for the psychic to receive information from his contact in the other world that he has no way of knowing ahead of time— say, the exact date of the future earthquake that levels Los Angeles.

Another commonly reported phenomenon that is used to claim evidence for an afterlife is the *near-death experience* (NDE). People very close to death who then survive often report seeing a tunnel with light at the end of it and someone beckoning to them in the light. Since the person was never brain-dead, she cannot be said to have come back from the dead. However, the claim is that she saw a sign of the world beyond at the end of the tunnel.

I provided an extensive critique of experiments on near-death experiences in *Has Science Found God?*[54] There we found that none provide any evidence for an afterlife. See also the book by Susan Blackmore.[55]

In a well-balanced assessment of the evidence of near-death experiences, *Religion, Spirituality, and the Near-Death Experience*, Mark Fox concludes: "This needs to be spelled out loudly and clearly: twenty-five years after the coining of the actual phrase 'near-death experience,' it remains to be established beyond doubt that during such an experience anything actually leaves the body. To date, and claims to the contrary notwithstanding, no researcher has provided evidence for such an assertion of an acceptable standard which would put the matter beyond doubt."[56]

In short, after over a century of unsuccessful attempts to find convincing scientific evidence for the almost universally desired immortal and immaterial soul, it seems very unlikely that it, and a God who provides us with such a gift, exists.

MODERN THEOLOGIES OF SOUL

Contemporary theologians are far from unaware that scientific developments in biology and neuroscience have undermined traditional beliefs about the soul and human nature. Theologian Nancey Murphy has written, "Science has provided a massive amount of evidence suggesting that we need not postulate the existence of an entity such as a soul or mind in order to account for life and consciousness."[57]

Murphy sees this as a serious theological problem, that Cartesian dualism is no longer tenable. She is certainly correct on that. However, she is unwilling to concede that the only option remaining is "reductive materialism," which she regards as incompatible with Christian teaching (not a scientific reason). Instead she has joined other theologians in proposing what she calls *nonreductive physicalism*. In this view, "The person is a physical organism whose complex functioning, both in society and in relation to God, gives rise to 'higher' human capacities such as morality and spirituality."[58]

Computer simulations of complex systems have uncovered a property that has provided Murphy and others what they think may be a scientific alternative to reductive materialism that has theological implications. These simulations have revealed unexpected features for systems as a whole that are not present in their various parts. This property is called *emergence* and is said to testify to a new holistic reality in which the whole is greater than the sum of its parts.

Psychologist Warren S. Brown suggests that the neurocognitive system has such emergent functions that cannot be reduced to "lower abilities," although he admits it would not exist without those lower abilities. Further, he claims, without evidence, that the human cognitive system has the ability of "downward causative influence" on those lower abilities.[59] Brown argues that the notion of "interpersonal interrelatedness" that emerges corresponds to the Christian experience of soul.[60]

If what emerges may be called the soul, it is still the product of purely material processes. Nothing supernatural is taking place, and God is an unnecessary ingredient. The wetness of water is an emergent property of H_2O molecules, but that doesn't imply the existence of some immaterial thing called wetness. Human and animal mental processes look just as they can be expected to look if there is no soul or other immaterial component.

As discussed above, physical processes display no properties that cannot be simply reduced to the localized interactions of its parts by well-known laws of physics that require some new "holistic" principles. Those properties follow from the same reducible physics as do the hardness of rock and the wetness of water.

In any case, whether reductive or not, the emergent properties of the purely physical brain and body do not survive their deaths. The nonreductive physicalist soul is not an immortal immaterial soul—not even a mortal immaterial soul. Once again it appears that a God with a traditional attribute of the monotheistic God, one who endows humans with immortal immaterial souls, does not exist.[61]

NOTES

1. Carl Zimmer, *Soul Made Flesh: The Discovery of the Brain—and How It Changed the World* (New York: Free Press, 2004), pp. 9–11.

2. Jerome W. Elbert, *Are Souls Real?* (Amherst, NY: Prometheus Books, 2000), p. 37; David Eller, *Natural Atheism* (Cranford, NJ: American Atheist Press, 2004), pp. 333–40.

3. Zimmer, *Soul Made Flesh*, p. 17.

4. Ibid. The Lucretius quotation is from Bernard Pullman, *The Atom in the History of Human Thought* (Oxford: Oxford University Press, 1998).

5. Ibid., p. 36.

6. Philip Ball, *Critical Mass: How One Thing Leads to Another* (New York: Farrar, Straus and Giroux, 2004), chap. 1.

7. Zimmer, *Soul Made Flesh*.

8. C. J. Aine, "A Conceptual Overview and Critique of Functional Neuro-Imaging Techniques in Humans: I. MRI/fMRI and PET," *Critical Reviews in Neurobiology* 9, nos. 2–3 (1995): 229–309.

9. Jorge Moll et al., "Functional Networks in Emotive and Non-moral Social Judgments," *NeuroImage* 16 (2002): 696–703.

10. Joshua D. Greene et al., "The Neural Bases of Cognitive Conflict and Control in Moral Judgment," *Neuron* 44 (2004): 389–400.

11. Michael A. Persinger, "Paranormal and Religious Beliefs May Be Mediated Differently by Subcortical and Cortical Phenomenological Process of the Temporal (Limbic) Lobes," *Perceptual and Motor Skills* 76 (1993): 247–51.

12. P. Granqvist et al., "Sensed Presence and Mystical Experiences Are Predicted by Suggestibility, Not by the Application of Transcranial Weak Complex Magnetic Fields," *Neuroscience Letters* 379, no. 1 (2005): 1–6.

13. Olaf Blanke et al., "Stimulating Illusory Own-Body Perceptions," *Nature* 419 (September 19, 2002): 269–70.

14. Victor J. Stenger, *Physics and Psychics: The Search for a World beyond the Senses* (Amherst, NY: Prometheus Books, 1990), p. 111; *Has Science Found God? The Latest Results in the Search for Purpose in the Universe* (Amherst, NY: Prometheus Books, 2003), pp. 290–99.

15. Pope Pius XII, *Humani Generis*, August 12, 1950.

16. Pope John Paul II, Address to the Academy of Sciences, October 28, 1986. *L'Osservatore Romano*, English ed., November 24, 1986, p. 22.

17. Stenger, *Physics and Psychics*.

18. Stenger, *Has Science Found God?*

19. Stenger, "Bioenergetic Fields," *Scientific Review of Alternative Medicine* 3, no. 1 (Spring/Summer 1999).

20. Joanne Stefanatos, "Introduction to Bioenergetic Medicine," in *Complementary and Alternative Veterinary Medicine: Principles and Practice*, ed. Allen M. Schoen and Susan G. Wynn (St. Louis: Mosby-Year Book, 1998), chap. 16.

21. L. Rosa et al., "A Close Look at Therapeutic Touch," *Journal of the American Medical Association* 279 (1998): 1005–10; Bela Scheiber and Carla Selby, eds., *Therapeutic Touch* (Amherst, NY: Prometheus Books, 2000).

22. Xin Yan et al., "Certain Physical Manifestation and Effects of External Qi of Yan Xin Life Science Technology," *Journal of Scientific Exploration* 16, no. 3 (2002): 381–411.

23. Stenger, *Physics and Psychics*.

24. See my discussion in Stenger, *Has Science Found God?* pp. 281–85.

25. Crookes was drawn into spiritualism after the death of his brother in 1867. Lodge's son was killed in Flanders in 1915, and Lodge turned to mediums to communicate with Raymond in the beyond.

26. Stenger, *Physics and Psychics*.

27. John P. A. Ionnidas, "Why Most Published Research Findings Are False," *Public Library of Science, Medicine* 2, no. 8 (2005), http://medicine.plosjournals.org/perlserv/?request=get-document&doi=10.1371/journal.pmed.0020124 (accessed December 2, 2005).

28. Jonathan A. Sterne and George Davey Smith, "Sifting the Evidence—What's Wrong with Significance Tests?" *British Medical Journal* 322 (2001): 226–31.

29. Stenger, *Physics and Psychics; Has Science Found God?*

30. Dean Radin, *The Conscious Universe: The Scientific Truth of Psychic Phenomena* (New York: HarperEdge, 1997).

31. Douglas M. Stokes, "The Shrinking Filedrawer: On the Validity

of Statistical Meta-Analysis in Parapsychology," *Skeptical Inquirer* 25, no. 3 (2001): 22–25.

32. Jeffrey P. Bishop and Victor J. Stenger, "Retroactive Prayer: Lots of History, Not Much Mystery, and No Science," *British Medical Journal* 329 (2004): 1444–46.

33. See, for example, Larry Dossey, *Healing Words: The Power of Prayer and the Practice of Medicine* (San Francisco: Harper, 1993).

34. Stenger, *Has Science Found God?* pp. 237–55.

35. K. Y. Cha, D. P. Wirth, and R. A. Lobo, "Does Prayer Influence the Success of In Vitro Fertilization-Embryo Transfer? Report of a Masked, Randomized Trial," *Journal of Reproductive Medicine* 46, no. 9 (September 2001): 781–87.

36. Timothy Johnson, "Praying for Pregnancy: Study Says Prayer Helps Women Get Pregnant," ABC Television, *Good Morning America*, October 4, 2001.

37. Bruce L. Flamm, "Faith Healing by Prayer," review of "Does Prayer Influence the Success of In Vitro Fertilization-Embryo Transfer?" by K. Y. Cha, D. P. Wirth, and R. A. Lobo, *Scientific Review of Alternative Medicine* 6, no. 1 (2002): 47–50; Bruce L. Flamm, "Faith Healing Confronts Modern Medicine," *Scientific Review of Alternative Medicine* 8, no. 1 (2004): 9–14.

38. See references in Flamm, "Faith Healing Confronts Modern Medicine."

39. Dossey, Response to letter to the editor, *Southern California Physician* (December 2001): 46.

40. Leonard Leibovici, "Effects of Remote, Retroactive Intercessory Prayer on Outcomes in Patients with Bloodstream Infections: A Controlled Trial," *British Medical Journal* 323 (2001): 1450–51.

41. Leonard Leibovici, "Alternative (Complementary) Medicine: A Cuckoo in the Nest of Empiricist Reed Warblers," *British Medical Journal* 319 (1999): 1629–31.

42. Brian Olshansky and Larry Dossey, "Retroactive Prayer: A Preposterous Hypothesis?" *British Medical Journal* 327 (2003): 1460–63.

43. Bishop and Stenger, "Retroactive Prayer."

44. Victor J. Stenger, *The Unconscious Quantum: Metaphysics in Modern Physics and Cosmology* (Amherst, NY: Prometheus Books, 1995).

45. Victor J. Stenger, *Timeless Reality: Symmetry, Simplicity, and Multiple Universes* (Amherst, NY: Prometheus Books, 2000).

46. Randolph C. Byrd, "Positive Therapeutic Effects of Intercessory Prayer in a Coronary Care Unit Population," *Southern Medical Journal* 81, no. 7 (1988): 826–29; W. S. Harris et al., "A Randomized, Controlled Trial of the Effects of Remote, Intercessory Prayer on Outcomes in Patients Admitted to the Coronary Care Unit," *Archives of Internal Medicine* 159 (1999): 2273–78.

47. M. W. Krucoff et al., "Music, Imagery, Touch, and Prayer as Adjuncts to Interventional Cardiac Care: The Monitoring and Actualization of Noetic Trainings (MANTRA) II Randomized Study," *Lancet* 366 (July 16, 2005): 211–17; for a media report, see Jonathan Petre, "Power of Prayer Found Wanting in Hospital Trial," *News Telegraph*, October 15, 2003, http://news.telegraph.co.uk/news/main.jhtml?xml=/news/2003/10/15/npray15.xml (accessed December 6, 2004).

48. Nathan Bupp, "Follow-up Study on Prayer Therapy May Help Refute False and Misleading Information about Earlier Prayer Study," *Commission for Scientific Medicine and Mental Health*, http://csmmh.org/prayer/MANTRA.release.htm, July 22, 2005 (accessed December 16, 2005).

49. Center for Spirituality, Theology and Health, http://www.duke spiritualityandhealth.org/books/ (accessed December 16, 2005).

50. H. Benson et al., "Study of the Therapeutic Effects of Intercessory Prayer (STEP) in Cardiac Bypass Patients: A Multicenter Randomized Trial of Uncertainty and Certainty of Receiving Intercessory Prayer," *American Heart Journal* 151, no. 4 (2006): 934–42.

51. 1 Corinthians 15:14.

52. Corliss Lamont, *The Illusion of Immortality*, 5th ed. (New York: Continuum, 1990). First published in 1935.

53. Ibid., pp. 43–44.

54. Stenger, *Has Science Found God?* pp. 290–99.

55. Susan Blackmore, *Dying to Live: Near-Death Experiences* (Amherst, NY: Prometheus Books, 1993).

56. Mark Fox, *Religion, Spirituality, and the Near-Death Experience* (New York: Routledge, 2003).

57. Nancey Murphy in *Whatever Happened to the Soul? Scientific and Theological Portraits of Human Nature*, ed. Warren S. Brown,

Nancey Murphy, and H. Newton Malony (Minneapolis: Fortress Press, 1998), p. 18.

58. Ibid., p. 25.

59. Warren S. Brown in *Whatever Happened to the Soul? Scientific and Theological Portraits of Human Nature*, ed. Warren S. Brown, Nancey Murphy, and H. Newton Malony (Minneapolis: Fortress Press, 1998), p. 102.

60. Ibid., p. 125.

61. For further reading on the philosophical aspects of the mind-body problem in the light of current research, see Daniel Dennett, *Consciousness Explained* (Boston: Little, Brown, 1991); Patricia Smith Churchland, *Neurophilosophy: Toward a Unified Science of the Mind/Brain* (Cambridge, MA: MIT Press, 1996); Paul M. Churchland, *The Engine of Reason, the Seat of the Soul: A Philosophical Journey into the Brain* (Cambridge, MA: MIT Press, 1996); George Lakoff and Mark Johnson, *Philosophy in the Flesh: The Embodied Mind and Its Challenge to Western Thought* (New York: Basic Books, 1999).

Chapter 4

COSMIC EVIDENCE

*The only laws of matter are those which our minds must fabricate,
and the only laws of mind are fabricated for it by matter.*
—James Clerk Maxwell

MIRACLES

Let us now move from Earth to the cosmos in our search for evidence of the creator God of Judaism, Christianity, and Islam. From a modern scientific perspective, what are the empirical and theoretical implications of the hypothesis of a supernatural creation? We need to seek evidence that the universe (1) had an origin and (2) that origin cannot have happened naturally. One sign of a supernatural creation would be a direct empirical confirmation that a miracle was necessary in order to bring the universe into existence. That is, cosmological data should either show evidence for one or more violations of well-established laws of nature or the models developed to describe those data should require some causal ingredient that cannot be understood—and be probably not understandable—in purely material or natural terms.

Now, as philosopher David Hume pointed out centuries ago, many problems exist with the whole notion of miracles. Three types of possible miracles can be identified: (1) violations of established laws of nature, (2) inexplicable events, and (3) highly unlikely coincidences. The latter two can be subsumed into the first since they also would imply a disagreement with current knowledge.

In previous chapters I have given examples of observations that would confirm the reality of supernatural powers of the human mind. We can easily imagine cosmic phenomena that would forever defy material expectations. Suppose a new planet were to suddenly appear in the solar system. Such an observation would violate energy conservation and reasonably be classified as a supernatural event.

Scientists will make every effort to find a natural mechanism for any unusual event, and the layperson is likely to agree that such a mechanism might be possible since "science does not know everything."

However, science knows a lot more than most people realize. Despite the talk of "scientific revolutions" and "paradigm shifts," the basic laws of physics are essentially the same today as they were at the time of Newton. Of course they have been expanded and revised, especially with the twentieth-century developments of relativity and quantum mechanics. But anyone familiar with modern physics will have to agree that certain fundamentals, in particular the great conservation principles of energy and momentum, have not changed in four hundred years.[1] The conservation principles and Newton's laws of motion still appear in relativity and quantum mechanics. Newton's law of gravity is still used to calculate the orbits of spacecraft.

Conservation of energy and other basic laws hold true in the most distant observed galaxy and in the cosmic microwave background, implying that these laws have been valid for over thirteen billion years. Surely any observation of their violation during the puny human life span would be reasonably termed a miracle.

Theologian Richard Swinburne suggests that we define a miracle as a nonrepeatable exception to a law of nature.[2] Of course, we can always redefine the law to include the exception, but that would be somewhat arbitrary. Laws are meant to describe repeatable events. So, we will seek evidence for violations of well-established laws that do not repeat themselves in any lawful pattern.

No doubt God, if he exists, is capable of repeating miracles if he so desires. However, repeatable events provide more information that may lead to an eventual natural description, while a mysterious, unrepeated event is likely to remain mysterious. Let us give the God hypothesis every benefit of the doubt and keep open the possibility of a miraculous origin for inexplicable events and unlikely coincidences, examining any such occurrences on an individual basis. If even with the loosest definition of a miracle none is observed to occur, then we will have obtained strong support for the case against the existence of a God who directs miraculous events.

Let us proceed to look for evidence of a miraculous creation in our observations of the cosmos.

CREATING MATTER

Until early in the twentieth century, there were strong indications that one or more miracles were required to create the universe. The universe currently contains a large amount of matter that is characterized by the physical quantity we define as mass. Prior to the twentieth century, it was believed that matter could neither be created nor destroyed, just changed from one type to another. So the very existence of matter seemed to be a miracle, a violation of the assumed law of conservation of mass that occurred just once—at the creation.

However, in his special theory of relativity published in 1905, Albert Einstein showed that matter can be created out of energy and can disappear into energy. What all science writers call "Einstein's famous equation," $E = mc^2$, relates the mass m of a body to an equivalent rest energy, E, where c is a universal constant, the speed of light in a vacuum. That is, a body at rest still contains energy.

When a body is moving, it carries an additional energy of motion called *kinetic energy*. In chemical and nuclear interactions, kinetic energy can be converted into rest energy, which is equivalent to generating mass.[3] Also, the reverse happens; mass or rest energy can be converted into kinetic energy. In that way, chemical and nuclear interactions can generate kinetic energy, which then can be used to run engines or blow things up.

So, the existence of mass in the universe violates no law of nature. Mass can come from energy. But, then, where does the energy come from? The law of conservation of energy, also known as the *first law of thermodynamics*, requires that energy come from somewhere. In principle, the creation hypothesis could be confirmed by the direct observation or theoretical requirement that conservation of energy was violated 13.7 billion years ago at the start of the big bang.

However, neither observations nor theory indicates this to have been the case. The first law allows energy to convert from one type to another as long as the total for a closed system remains fixed. Remarkably, the total energy of the universe appears to be zero. As famed cosmologist Stephen Hawking said in his 1988 best seller, *A Brief History of Time*, "In the case of a universe that is approximately uniform in space, one can show that the negative gravitational energy exactly cancels the positive energy represented by the matter. So the total energy of the universe is zero.[4] Specifically, within small measurement errors, the mean energy density of the universe is exactly what it should be for a universe that appeared from an initial state of zero energy, within a small quantum uncertainty.[5]

A close balance between positive and negative energy is predicted by the modern extension of the big bang theory called the *inflationary big bang*, according to which the universe underwent a period of rapid, exponential inflation during a tiny fraction of its first second.[6] The inflationary theory has recently undergone a number of stringent observational tests that would have been sufficient to prove it false. So far, it has successfully passed all these tests.

In short, the existence of matter and energy in the universe did not require the violation of energy conservation at the assumed creation. In fact, the data strongly support the hypothesis that no such miracle occurred. If we regard such a miracle as predicted by the creator hypothesis, then that prediction is not confirmed.

This example also serves to once more refute the assertion that science has nothing to say about God. Suppose our measurement of the mass density of the universe had *not* turned out to be exactly the value required for a universe to have begun from a state of zero energy. Then we would have had a legitimate, scientific reason to conclude that a miracle, namely, a violation of energy conservation, was needed to bring the universe into being. While this might not conclusively prove the existence of a creator to everyone's satisfaction, it would certainly be a strong mark in his favor.

CREATING ORDER

Another prediction of the creator hypothesis also fails to be confirmed by the data. If the universe were created, then it should have possessed some degree of order at the creation—the design that was inserted at that point by the Grand Designer. This expectation of order is usually expressed in terms of the *second law of thermodynamics*, which states that the total *entropy* or *disorder* of a closed system must remain constant or increase with time. It would seem to follow that if the universe today is a closed system, it could not always have been so. At some point in the past, order must have been imparted from the outside.

Prior to 1929, this was a strong argument for a miraculous creation. However, in that year astronomer Edwin Hubble reported that the galaxies are moving away from one another at speeds approximately proportional to their distance, indicating that the universe is expanding. This provided the earliest evidence for the big bang. For our purposes, an expanding universe could have started in total chaos and still formed localized order consistent with the second law.

The simplest way to see this is with a (literally) homey example. Suppose that whenever you clean your house, you empty the collected rubbish by tossing it out the window into your yard. Eventually the yard would be filled with rubbish. However, you can continue doing this with a simple expedient. Just keep buying up the land around your house and you will always have more room to toss the rubbish. You are able to maintain localized order—in your house—at the expense of increased disorder in the rest of the universe.

Similarly, parts of the universe can become more orderly as the rubbish, or entropy, produced during the ordering process (think of it as disorder being removed from the system being ordered) is tossed out into the larger, ever-expanding surrounding space. As illustrated in figure 4.1, the total entropy of the universe increases as the universe expands, as required by the second law.[7] However, the maximum possible entropy increases even faster, leaving increasingly more room for order to form. The reason for this is that the maximum entropy of a sphere of a certain radius (we are thinking of the universe as a sphere) is that of a black hole of that radius. The expanding universe is not a

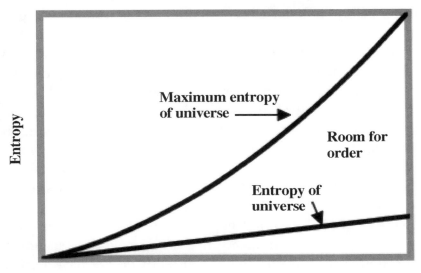

Maximum entropy of universe ⟶

Room for order

Entropy of universe

Planck time

Radius of the universe

Fig. 4.1. The total entropy of the universe and the maximum entropy as a function of the radius of the universe. They are equal at the origin, the Planck time, which shows that the universe begins in total chaos. However, since the universe is expanding, the maximum entropy increases faster than the actual total entropy leaving increasing room for order to form without violating the second law of thermodynamics.

black hole and so has less than maximum entropy. Thus, while becoming more disorderly on the whole as time goes by, our expanding universe is not maximally disordered. But, once it was.

Suppose we extrapolate the expansion back 13.7 billion years to the earliest definable moment, the *Planck time*, 6.4×10^{-44} second when the universe was confined to the smallest possible region of space that can be operationally defined, a *Planck sphere* that has a radius equal to the *Planck length*, 1.6×10^{-35} meter. As expected from the second law, the universe at that time had lower entropy than it has now. However, that entropy was also as high as it possibly could have been for an object that small, because a sphere of Planck dimensions is equivalent to a black hole.

This requires further elaboration. I seem to be saying that the entropy of the universe was maximal when the universe began, yet it has been increasing ever since. Indeed, that's exactly what I am saying. When the universe began, its entropy was as high as it could be for an object of that size because the universe was equivalent to a black hole from which no information can be extracted. Currently, the entropy is higher but not maximal, that is, not as high as it could be for an object of the universe's current size. The universe is no longer a black hole.

I also need to respond here to an objection that has been raised by physicists who have heard me make this statement. They point out, correctly, that we currently do not have a theory of quantum gravity that we can apply to describe physics earlier than the Planck time. I have adopted Einstein's operational definition of time as what you read on a clock. In order to measure a time interval smaller than the Planck time, you would need to make that measurement in a region smaller than the Planck length, which equals the Planck time multiplied by the speed of light. According to the Heisenberg uncertainty principle of quantum mechanics, such a region would be a black hole, from which no information can escape. This implies that no time interval can be defined that is smaller than the Planck time.[8]

Consider the present time. Clearly we do not have any qualms about applying established physics "now" and for short times earlier or later, as long as we do not try to do so for time intervals shorter than the Planck time. Basically, by definition time is counted off as an integral number of units where one unit equals the Planck time. We can get away with treating time as a continuous variable in our mathematical physics, such as we do when we use calculus, because the units are so small compared to anything we measure in practice. We essentially extrapolate our equations through the Planck intervals within which time is unmeasurable and thus indefinable. If we can do this "now," we can do it at the end of the earliest Planck interval where we must begin our description of the beginning of the big bang.

At that time, our extrapolation from later times tells us that the entropy was maximal. In that case, the disorder was complete and no structure could have been present. Thus, the universe began with no structure. It has structure today consistent with the fact that its entropy is no longer maximal.

In short, according to our best current cosmological understanding, our universe began with no structure or organization, designed or otherwise. It was a state of chaos.

We are thus forced to conclude that the complex order we now observe could *not* have been the result of any initial design built into the universe at the so-called creation. The universe preserves no record of what went on before the big bang. The Creator, if he existed, left no imprint. Thus he might as well have been nonexistent.

Once again we have a result that might have turned out otherwise and provided strong scientific evidence for a creator. If the universe were not expanding but a firmament, as described in the Bible, then the second law would have required that the entropy of the universe was lower than its maximum allowed value in the past. Thus, if the universe had a beginning, it would have begun in a state of high order necessarily imposed from the outside. Even if the universe extended into the infinite past, it would be increasingly orderly in that direction, and the source of that order would defy natural description.

BEGINNING AND CAUSE

The empirical fact of the big bang has led some theists to argue that this, in itself, demonstrates the existence of a creator. In 1951 Pope Pius XII told the Pontifical Academy, "Creation took place in time, therefore there is a Creator, therefore God exists."[9] The astronomer/priest Georges-Henri Lemaître, who first proposed the idea of a big bang, wisely advised the pope not make this statement "infallible."

Christian apologist William Lane Craig has made a number of sophisticated arguments that he claims show that the universe must have had a beginning and that beginning implies a personal creator.[10] One such argument is based on *general relativity*, the modern theory of gravity that was published by Einstein in 1916 and that has, since then, passed many stringent empirical tests.[11]

In 1970 cosmologist Stephen Hawking and mathematician Roger Penrose, using a theorem derived earlier by Penrose, "proved" that a *singularity* exists at the beginning of the big bang.[12] Extrapolating general relativity back to zero time, the universe gets smaller and smaller while the density of the universe and the gravitational field increases. As the size of the universe goes to zero, the density and gravitational field, at least according to the mathematics of general relativity, become infinite. At that point, Craig claims, time must stop and, therefore, no prior time can exist.

However, Hawking has repudiated his own earlier proof. In his best seller *A Brief History of Time*, he avers, "There was in fact no singularity at the beginning of the universe."[13] This revised conclusion, concurred with by Penrose, follows from quantum mechanics, the theory of atomic processes that was developed in the years following the introduction of Einstein's theories of relativity. Quantum mechanics, which also is now confirmed to great precision, tells us that general relativity, at least as currently formulated, must break down at times less than the Planck time and at distances smaller than the Planck length, mentioned earlier. It follows that general relativity cannot be used to imply that a singularity occurred prior to the Planck time and that Craig's use of the singularity theorem for a beginning of time is invalid.

Craig and other theists also make another, related argument that the universe had to have had a beginning at some point, because if it were infinitely old, it would have taken an infinite time to reach the present. However, as philosopher Keith Parsons has pointed out, "To say the universe is infinitely old is to say that it had no beginning—not a beginning that was infinitely long ago."[14]

Infinity is an abstract mathematical concept that was precisely formulated in the work of mathematician Georg Cantor in the late nineteenth century. However, the symbol for infinity, "∞," is used in physics simply as a shorthand for "a very big number." Physics is counting. In physics, time is simply the count of ticks on a clock. You can count backward as well as forward. Counting forward you can get a very big but never mathematically infinite positive number and time "never ends." Counting backward you can get a very big but never mathematically infinite negative number and time "never begins." Just as we never reach positive infinity, we never reach negative infinity. Even if the universe does not have a mathematically infinite number of events in the future, it still need not have an end. Similarly, even if the universe does not have a mathematically infinite number of events in the past, it still need not have a beginning. We can always have one event follow another, and we can always have one event precede another.

Craig claims that if it can be shown that the universe had a beginning, this is sufficient to demonstrate the existence of a personal creator. He casts this in terms of the *kalâm cosmological argument*, which is drawn from Islamic theology.[15] The argument is posed as a syllogism:

1. Whatever begins to exist has a cause.
2. The universe began to exist.
3. Therefore, the universe has a cause.

The *kalâm* argument has been severely challenged by philosophers on logical grounds,[16] which need not be repeated here since we are focusing on the science.

In his writings, Craig takes the first premise to be self-evident, with no justification other than common, everyday experience. That's the type of experience that tells us the world is flat. In fact, physical events at the atomic and subatomic level are observed to have no evident cause. For example, when an atom in an excited energy level drops to a lower level and emits a photon, a particle of light, we find no cause of that event. Similarly, no cause is evident in the decay of a radioactive nucleus.

Craig has retorted that quantum events are still "caused," just caused in a nonpredetermined manner—what he calls "probabilistic causality." In effect, Craig is thereby admitting that the "cause" in his first premise could be an accidental one, something spontaneous—something not predetermined. By allowing probabilistic cause, he destroys his own case for a predetermined creation.

We have a highly successful theory of probabilistic causes—quantum mechanics. It does not predict when a given event will occur and, indeed, assumes that individual events are not predetermined. The one exception occurs in the interpretation of quantum mechanics given by David Bohm.[17] This assumes the existence of yet-undetected subquantum forces. While this interpretation has some supporters, it is not generally accepted because it requires superluminal connections that violate the principles of special relativity.[18] More important, no evidence for subquantum forces has been found.

Instead of predicting individual events, quantum mechanics is used to predict the statistical distribution of outcomes of ensembles of similar events. This it can do with high precision. For example, a quantum calculation will tell you how many nuclei in a large sample will have decayed after a given time. Or you can predict the intensity of light from a group of excited atoms, which is a measure of the total number of photons emitted. But neither quantum mechanics nor any other existing theory—including Bohm's—can say anything about the behavior of an individual nucleus or atom. The photons emitted in atomic transitions come into existence spontaneously, as do the particles emitted in nuclear radiation. By so appearing, without predetermination, they contradict the first premise.

In the case of radioactivity, the decays are observed to follow an exponential decay "law." However, this statistical law is exactly what you expect if the probability for decay in a given small time interval is the same for all time intervals of the same duration. In other words, the decay curve itself is evidence for each individual event occurring unpredictably and, by inference, without being predetermined.

Quantum mechanics and classical (Newtonian) mechanics are not as separate and distinct from one another as is generally thought. Indeed, quantum mechanics changes smoothly into classical mechanics when the parameters of the system, such as masses, distances, and speeds, approach the classical regime.[19] When that happens, quantum probabilities collapse to either zero or 100 percent, which then gives us certainty at that level. However, we have many examples where the probabilities are not zero or 100 percent. The quantum probability calculations agree precisely with the observations made on ensembles of similar events.

Note that even if the *kalâm* conclusion were sound and the universe had a cause, why could that cause itself not be natural? As it is, the *kalâm* argument fails both empirically and theoretically without ever having to bring up the second premise about the universe having a beginning.

THE ORIGIN

Nevertheless, another nail in the coffin of the *kalâm* argument is provided by the fact that the second premise also fails. As we saw above, the claim that the universe began with the big bang has no basis in current physical and cosmological knowledge.

The observations confirming the big bang do not rule out the possibility of a prior universe. Theoretical models have been published suggesting mechanisms by which our current universe appeared from a preexisting one, for example, by a process called quantum tunneling or so-called quantum fluctuations.[20] The equations of cosmology that describe the early universe apply equally for the other side of the time axis, so we have no reason to assume that the universe began with the big bang.

In *The Comprehensible Cosmos*, I presented a specific scenario for the purely natural origin of the universe, worked out mathematically at a level accessible to anyone with an undergraduate mathematics or physics background.[21] This was based on the *no boundary model* of James Hartle and Stephen Hawking.[22] In that model, the universe has no beginning or end in space or time. In the scenario I presented, our universe is described as having "tunneled" through the chaos at the Planck time from a prior universe that existed for all previous time.

While he avoided technical details in *A Brief History of Time*, the no boundary model was the basis of Hawking's oft-quoted statement: "So long as the universe had a beginning, we could suppose it had a creator. But if the universe is really completely self-contained, having no boundary or edge, it would have neither beginning nor end; it would simply be. What place then, for a creator?"[23]

Prominent physicists and cosmologists have published, in reputable scientific journals, a number of other scenarios by which the universe could have come about "from nothing" naturally.[24] None can be "proved" at this time to represent the exact way the universe appeared, but they serve to illustrate that any argument for the existence of God based on this gap in scientific knowledge fails, since plausible natural mechanisms can be given within the framework of existing knowledge.

As I have emphasized, the God of the gaps argument for God fails when a plausible scientific account for a gap in current knowledge can be given. I do not dispute that the exact nature of the origin of the universe remains a gap in scientific knowledge.

But I deny that we are bereft of any conceivable way to account for that origin scientifically.

In short, empirical data and the theories that successfully describe those data indicate that the universe did not come about by a purposeful creation. Based on our best current scientific knowledge, it follows that no creator exists who left a cosmological imprint of a purposeful creation.

INTERVENING IN THE COSMOS

This still leaves open the possibility that a god exists who may have created the universe in such a way that did not require any miracles and did not leave any imprint of his intentions. Of course, this is no longer the traditional Judeo-Christian-Islamic God, whose imprint is supposedly everywhere. But, perhaps those religions can modify their theologies and posit a god who steps in later, after the Planck time, to ensure that his purposes are still served despite whatever plans he had of creation being wiped out by the chaos at the Planck time.

In that case, we can again expect to find, in observations or well-established theories, some evidence of places where this god has intervened in the history of the cosmos. In previous chapters we sought such evidence on Earth, in the phenomena of life and mind. Here we move to the vast space beyond Earth.

History gives us many examples of unexpected events in the heavens that at first appeared miraculous. In 585 BCE a total eclipse of the sun over Asia Minor ended a battle between the Medes and the Lydians, with both sides fleeing in terror. In probably the first known case of a scientific prediction, Thales of Miletus had predicted the eclipse based on Babylonian records.

Eclipses are sufficiently rare that they are not so regular a part of normal human experience as are the rising and setting of the sun and the phases of the moon. However, they do repeat and behave lawfully, as do these more familiar phenomena. That's why today we can give the exact date (on our current calendar) of Thales's eclipse: May 28, 585 BCE. This demonstrates the remarkable power of science to both predict the future and postdict the past. About that time, Nebuchadnezzar II destroyed Jerusalem and carried the Judeans off into exile in Babylonia (where they would pick up their creation myth). The Buddha is said to have attained enlightenment at almost exactly the same time. Confucius would be born a few decades later.

Comets are a similar example of spectacular astronomical phenomena that ancient people commonly regarded as supernatural omens but science has since described in natural terms, that is, with purely material models. In the seventeenth century, Edmund Halley (d. 1742) used the mechanical theories developed by his friend Isaac Newton (d. 1727) to predict that a comet seen in 1682 would return in 1759. Indeed it did, after Halley's death, and has done so every seventy-six years since. Most comets appear unexpectedly, having such extended orbits that they have spent human history outside our view. However, records indicate that Halley's comet has appeared perhaps twenty-nine times in history.

In more recent times, other astronomical phenomena have occurred unexpectedly and could not be immediately understood. These include pulsars, supernovas, quasars, and gamma-ray bursts. But, as with other examples, these phenomena eventually repeated in one way or another, in time or in space. This allowed us to learn enough to eventually understand their nature in purely physical terms.

At no time and at no place in the sky have we run across an event above the noise that did not repeat sometime or someplace and could not be accounted for in terms of established natural science. We have yet to encounter an observable astronomical phenomenon that requires a supernatural element to be added to a model in order to describe the event. In fact, we have no cosmic phenomenon that meets the Swinburne criterion for a miracle. A

God who plays a sufficiently active role to produce miraculous events in the cosmos has not been even glimpsed at by our best astronomical instruments to date. Observations in cosmology look just as they can be expected to look if there is no God.

WHERE DO THE LAWS OF PHYSICS COME FROM?

We have seen that the origin and the operation of the universe do not require any violations of laws of physics. This probably will come as a surprise to the layperson who may have heard otherwise from the pulpit or the media. However, the scientifically savvy believer might concede this point for the sake of argument and then retort, "Okay, then where did the laws of physics come from?" The common belief is that they had to come from somewhere outside the universe. But that is not a demonstrable fact. There is no reason why the laws of physics cannot have come from within the universe itself.

Physicists invent mathematical models to describe their observations of the world. These models contain certain general principles that have been traditionally called "laws" because of the common belief that these are rules that actually govern the universe the way civil laws govern nations. However, as I showed in my previous book, *The Comprehensible Cosmos*, the most fundamental laws of physics are not restrictions on the behavior of matter. Rather they are restrictions on the way physicists may describe that behavior.[25]

In order for any principle of nature we write down to be objective and universal, it must be formulated in such a way that it does not depend on the point of view of any particular observer. The principle must be true for all point of views, from every "frame of reference." And so, for example, no objective law can depend on a special moment in time or a position in space that may be singled out by some preferred observer.

Suppose I were to formulate a law that said that all objects move naturally toward me. That would not be very objective. But this was precisely what people once thought—that Earth was the center of the universe and the natural motion of bodies was toward Earth. The Copernican revolution showed this was wrong and was the first step in the gradual realization of scientists that their laws must not depend on frame of reference.

In 1918 mathematician Emmy Noether proved that the most important physical laws of all—conservation of energy, linear momentum, and angular momentum—will automatically appear in any model that does not single out a special moment in time, position in space, and direction in space.[26] Later it was realized that Einstein's special theory of relativity follows if we do not single out any special direction in four-dimensional space-time.

These properties of space-time are called *symmetries*. For example, the rotational symmetry of a sphere is a result of the sphere singling out no particular direction in space. The four space-time symmetries described above are just the natural symmetries of a universe with no matter, that is, a void. They are just what they should be if the universe appeared from an initial state in which there was no matter—from nothing.

Other laws of physics, such as conservation of electric charge and the various force laws, arise from the generalization of space-time symmetries to the abstract spaces physicists use in their mathematic models. This generalization is called *gauge invariance*, which is likened to a principle I more descriptively refer to as *point-of-view invariance*.

The mathematical formulations of these models (which are provided in *The Comprehensible Cosmos*) must reflect this requirement if they are to be objective and universal. Surprisingly, when this is done, most of the familiar laws of physics appear naturally. Those that are not immediately obvious can be seen to plausibly arise by a process, mentioned in chapter 2, known as *spontaneous symmetry breaking*.

So where did the laws of physics come from? They came from nothing! Most are statements composed by humans that follow from the symmetries of the void out of which the universe spontaneously arose. Rather than being handed down from above, like the Ten Commandments, they look exactly as they should look if they were not handed down from anywhere. And this is why, for example, a violation of energy conservation at the beginning of the big bang would be evidence for some external creator. Even though they invented it, physicists could not simply change the "law." It would imply a miracle or, more explicitly, some external agency that acted to break the time symmetry that leads to conservation of energy. But, as we have seen, no such miracle is required by the data.

Thus we are justified in applying the conservation laws to the beginning of the big bang at the Planck time. At that time, as we saw earlier in this chapter, the universe had no structure. That meant that it had no distinguishable place, direction, or time. In such a situation, the conservation laws apply.

Now, this is certainly not a commonly understood view. Normally we think of laws of physics as part of the structure of the universe. But here I am arguing that the three great conservation laws are not part of any structure. Rather they follow from the very lack of structure at the earliest moment.

No doubt this concept is difficult to grasp. My views on this particular issue are not recognized by a consensus of physicists, although I insist that the science I have used is well established and conventional. I am proposing no new physics or cosmology but merely providing an interpretation of established knowledge in those fields as it bears on the question of the origin of physical law, a question few physicists ever ponder.

I must emphasize another important point, which has been frequently misunderstood. I am not suggesting that the laws of physics can be anything we want them to be, that they are merely "cultural narratives," as has been suggested by authors associated with the movement called postmodernism.[27] They are what they are because they agree with the data.

Whether or not you will buy into my account of the origin of physical law, I hope you will allow that I have at minimum provided a plausible natural scenario for a gap in scientific knowledge, that gap being a clear consensus on the origin of physical law. Once again, I do not have the burden of proving this scenario. The believer who wishes to argue that God is the source of physical law has the burden of proving (1) that my account is wrong, (2) that no other natural account is possible, and (3) that God did it.

WHY IS THERE SOMETHING RATHER THAN NOTHING?

If the laws of physics follow naturally from empty space-time, then where did that empty space-time come from? Why is there something rather than nothing? This question is often the last recourse of the theist who seeks to argue for the existence of God from physics and cosmology and finds that all his other arguments fail. Philosopher Bede Rundle calls it "philosophy's central, and most perplexing, question." His simple (but booklength) answer: "There has to be something."[28]

Clearly many conceptual problems are associated with this question. How do we define "nothing"? What are its properties? If it has properties, doesn't that make it something? The theist claims that God is the answer. But, then, why is there God rather than nothing? Assuming we can define "nothing," why should nothing be a more natural state of affairs than something? In fact, we can give a plausible scientific reason based on our best current knowledge of physics and cosmology that something is more natural than nothing!

In chapter 2 we saw how nature is capable of building complex structures by processes of self-organization, how simplicity begets complexity. Consider the example of the snowflake, the beautiful six-pointed pattern of ice crystals that results from the direct freezing of water vapor in the atmosphere. Our experience tells us that a snowflake is very ephemeral, melting quickly into drops of liquid water that exhibit far less structure. But that is only because we live in a relatively high-temperature environment, where heat reduces the fragile arrangement of crystals to a simpler liquid. Energy is required to break the symmetry of a snowflake.

In an environment where the ambient temperature is well below the melting point of ice, as it is in most of the universe far from the highly localized effects of stellar heating, any water vapor would readily crystallize into complex, asymmetric structures. Snowflakes would be eternal, or at least would remain intact until cosmic rays tore them apart.

This example illustrates that many simple systems of particles are unstable, that is, have limited lifetimes as they undergo spontaneous phase transitions to more complex structures of lower energy. Since "nothing" is as simple as it gets, we cannot expect it to be very stable. It would likely undergo a spontaneous phase transition to something more complicated, like a universe containing matter. The transition of nothing-to-something is a natural one, not requiring any agent. As Nobel laureate physicist Frank Wilczek has put it, "The answer to the ancient question 'Why is there something rather than nothing?' would then be that 'nothing' is unstable."[29]

In the nonboundary scenario for the natural origin of the universe I mentioned earlier, the probability for there being something rather than nothing actually can be calculated; it is over 60 percent.[30]

In short, the natural state of affairs is something rather than nothing. An empty universe requires supernatural intervention—not a full one. Only by the constant action of an agent outside the universe, such as God, could a state of nothingness be maintained. The fact that we have something is just what we would expect if there is no God.

NOTES

1. Conservation of energy was not immediately recognized but was already implicit in Newton's laws of mechanics.

2. Richard Swinburne, *The Existence of God* (Oxford: Clarendon Press, 1979), p. 229.

3. It is commonly thought that only nuclear reactions convert between rest and kinetic energy. This also happens in chemical reactions. However, the changes in the masses of the reactants in that case are too small to be generally noticed.

4. Stephen W. Hawking, *A Brief History of Time: From the Big Bang to Black Holes* (New York: Bantam, 1988), p. 129.

5. Technically, the total energy of the universe cannot be defined for all possible situations in general relativity. However, in V. Faraoni and F. I. Cooperstock, "On the Total Energy of Open Friedmann-Robertson-Walker Universes," *Astrophysical Journal* 587 (2003): 483–86, it is shown that the total energy of the universe can be defined for the most common types of cosmologies and is zero in these cases. This includes the case where the density is critical.

6. Alan Guth, *The Inflationary Universe* (New York: Addison-Wesley, 1997).

7. The mathematical derivation of the curves on this plot is given in appendix C of Victor J. Stenger, *Has Science Found God? The Latest Results in the Search for Purpose in the Universe* (Amherst, NY: Prometheus Books, 2003), pp. 356–57.

8. The mathematical proof of this is given in appendix A, Stenger, *Has Science Found God?* pp. 351–53.

9. Pope Pius XII, "The Proofs for the Existence of God in the Light of Modern Natural Science," Address by Pope Pius XII to the Pontifical Academy of Sciences, November 22, 1951, reprinted as "Modern Science and the Existence of God," *Catholic Mind* 49 (1972): 182–92.

10. William Lane Craig and Quentin Smith, *Theism, Atheism, and Big Bang Cosmology* (Oxford: Clarendon Press, 1997).

11. Clifford M. Will, *Was Einstein Right? Putting General Relativity to the Test* (New York: Basic Books, 1986).

12. Stephen W. Hawking and Roger Penrose, "The Singularities of

Gravitational Collapse and Cosmology," *Proceedings of the Royal Society of London*, series A, 314 (1970): 529–48.

13. Hawking, *A Brief History of Time*, p. 50.

14. Keith Parsons, "Is There a Case for Christian Theism?" in *Does God Exist? The Debate between Theists & Atheists*, J. P. Moreland and Kai Nielsen (Amherst, NY: Prometheus Books, 1993), p. 177. See also Wes Morriston, "Creation *Ex Nihilo* and the Big Bang," *Philo* 5, no. 1 (2002): 23–33.

15. William Lane Craig, *The Kalâm Cosmological Argument*, Library of Philosophy and Religion (London: Macmillan, 1979); *The Cosmological Argument from Plato to Leibniz*, Library of Philosophy and Religion (London: Macmillan, 1980).

16. Smith in *Theism, Atheism, and Big Bang Cosmology*, by Craig and Smith; Graham Oppy, "Arguing *About* The *Kalam* Cosmological Argument," *Philo* 5, no. 1 (Spring/Summer 2002): 34–61, and references therein; Arnold Guminski, "The Kalam Cosmological Argument: The Questions of the Metaphysical Possibility of an Infinite Set of Real Entities," *Philo* 5, no. 2 (Fall/Winter 2002): 196–215; Nicholas Everitt, *The Non-Existence of God* (London, New York: Routledge, 2004), pp. 68–72.

17. David Bohm and B. J. Hiley, *The Undivided Universe: An Ontological Interpretation of Quantum Mechanics* (London: Routledge, 1993).

18. I discuss this in detail in Victor J. Stenger, *The Unconscious Quantum: Metaphysics in Modern Physics and Cosmology* (Amherst, NY: Prometheus Books, 1995).

19. Quantum mechanics becomes classical mechanics when Planck's constant h is set equal to zero.

20. David Atkatz and Heinz Pagels, "Origin of the Universe as Quantum Tunneling Event," *Physical Review* D25 (1982): 2065–67; Alexander Vilenkin, "Birth of Inflationary Universes," *Physical Review* D27 (1983): 2848–55; David Atkatz, "Quantum Cosmology for Pedestrians," *American Journal of Physics* 62 (1994): 619–27.

21. Victor J. Stenger, *The Comprehensible Cosmos: Where Do the Laws of Physics Come From?* (Amherst, NY: Prometheus Books, 2006), supplement H.

22. J. B. Hartle and S. W. Hawking, "Wave Function of the Universe," *Physical Review* D28 (1983): 2960–75.

23. Hawking, *A Brief History of Time*, pp. 140–41.

24. E. P. Tryon, "Is the Universe a Quantum Fluctuation?" *Nature* 246 (1973): 396–97; Atkatz and Pagels, "Origin of the Universe as Quantum Tunneling Event"; Alexander Vilenkin, "Quantum Creation of Universes," *Physical Review* D30 (1984): 509; Andre Linde, "Quantum Creation of the Inflationary Universe," *Lettere Al Nuovo Cimento* 39 (1984): 401–405; T. R. Mongan, "Simple Quantum Cosmology: Vacuum Energy and Initial State," *General Relativity and Gravitation* 37 (2005): 967–70.

25. Stenger, *The Comprehensible Cosmos*.

26. E. Noether, "Invarianten beliebiger Differentialausdrücke," *Nachr. d. König. Gesellsch. d. Wiss. zu Göttingen, Math-phys.* Klasse (1918): 37–44; Nina Byers, "E. Noether's Discovery of the Deep Connection between Symmetries and Conservation Laws," *Israel Mathematical Conference Proceedings* 12 (1999), http://www.physics.ucla.edu/~cwp/articles/noether.asg/noether.html (accessed July 1, 2006). This contains links to Noether's original paper including an English translation.

27. Walter Truett Anderson, *The Truth about the Truth* (New York: Jeremy P. Tarcher/Putnam, 1996).

28. Bede Rundle, *Why There Is Something Rather Than Nothing* (Oxford: Clarendon Press, 2004).

29. Frank Wilczek, "The Cosmic Asymmetry between Matter and Antimatter," *Scientific American* 243, no. 6 (1980): 82–90.

30. Stenger, *The Comprehensible Cosmos*, supplement H.

Chapter 5

THE UNCONGENIAL UNIVERSE

There can be no demonstrative argument to prove that those instances in which we have no experience, resemble those of which we have had experience.

—David Hume

THE PRIVILEGED PLANET

Human life is very sensitive to the physical conditions on Earth. If the atmosphere were not transparent to light in the so-called visible region of the electromagnetic spectrum, and if the sun did not provide light in that region, then our eyes would not be of any use. But, does this mean that the sun and Earth were specifically designed with those properties because human eyes are sensitive to the visible spectrum of light? As silly as that suggestion sounds, we hear similar arguments today presented as evidence for intelligent design in the universe. Of course the arguments are not presented in exactly that fashion but coated in a veneer of scientific-sounding language. But when that thin veneer is ripped away we are left with the even thinner substance underneath.

In his 1995 book, *The Creator and the Cosmos*, physicist Hugh Ross listed thirty-three characteristics a planet must have to support life. He also estimated the probability that such a combination be found in the universe as "much less than one in a million trillion."[1] He concluded that only "divine design" could account for human life.

However, Ross presented no estimate of the probability for divine design. Perhaps it is even lower! Ross and others who attempt to prove the existence of God on the basis of probabilities make a fundamental logical error. When using probabilities to decide between two or more possibilities, you must have a number for each possibility in order to compare. In this vast universe, highly unlikely events happen every day.

In a 2004 book called *The Privileged Planet*, astronomer Guillermo Gonzalez and theologian Jay Richards have carried the notion further, asserting that our place in the cosmos is not only special but also designed for discovery. They contend that conditions on Earth, particularly those that make human life possible, are also optimized for scientific investigation and that this constitutes "a signal revealing a universe so skillfully created for life and discovery that it seems to whisper of an extraterrestrial intelligence immeasurably more vast, more ancient, and more magnificent than anything we've been willing to expect or imagine."[2] Oh, come on, guys, you are willing to imagine who that intelligence is.

Following this line of reasoning, the atmosphere of Earth is not only transparent in the visible spectral band so that humans can see with their eyes, but it also is designed in this way so that astronomers can build telescopes and thereby observe the fruits of divine creation in the heavens.

Have you ever wondered why the angular diameters of the moon and sun as viewed from Earth are almost exactly the same, though the two celestial objects differ greatly in size and distance from Earth? Without that coincidence, we would never experience the type of total eclipse of the sun in which we can actually view starlight near the edge of the sun's disk.

Gonzalez and Richards marvel at the fact that we happen to live on a planet where total solar eclipses are observable, and present this as an example of design for discovery. As we saw in chapter 4, in 585 BCE Thales of Miletus predicted a total eclipse that supposedly ended a war. In more recent times, observations made during total eclipses have been used to verify Einstein's theory of general relativity, specifically the bending of starlight near the sun's edge. Gonzalez and Richards seem to think general relativity would not have been discovered (assuming the theories of physics are "out there" to be discovered, a notion I disputed in the previous chapter) had we lived on a planet without the coincidence of angular diameters. That is very dubious, since many other tests of general relativity have been made that do not involve eclipses.[3]

The privileged planet argument is reminiscent of the proposal by eighteenth-century German philosopher Gottfried Wilhelm Leibniz (d. 1716) that we live "in the best of all possible worlds." Leibniz was one of the greatest thinkers of all time, the independent coinventor (with Newton) of calculus. But this particular notion was ridiculed by the French philosopher François-Marie Arouet de Voltaire (d. 1778) in his short story "Candide." There, Dr. Pangloss, a thinly disguised Leibniz, proclaims:

It is demonstrable that things cannot be otherwise than as they are; for as all things have been created for some end, they must necessarily be created for the best end. Observe, for instance, the nose is formed for spectacles, therefore we wear spectacles. The legs are visibly designed for stockings, accordingly we wear stockings. Stones were made to be hewn and to construct castles, therefore My Lord has a magnificent castle; for the greatest baron in the province ought to be the best lodged. Swine were intended to be eaten, therefore we eat pork all the year round: and they, who assert that everything is right, do not express themselves correctly; they should say that everything is best.[4]

Gonzalez and Richards are senior fellows of the Center for Science and Culture,[5] the arm of the Seattle-based Discovery Institute that, as we saw earlier, is charged with the task of bringing science and culture into line with evangelical Christian teachings by driving "wedges" between materialistic science and the rest of society.[6]

The Privileged Planet constitutes a new wedge, a form of intelligent design designed to split conventional astronomy and physics off from the mainstream of public awareness. In 2005 the Discovery Institute produced a slick film under the same title presenting the arguments from the book. As with intelligent design in biology, the sectarian motives of the book and film were kept well hidden. So, when the film was presented by the Discovery Institute to the Smithsonian Institution for a special showing at the National Museum of Natural History in Washington, DC, along with a $16,000 fee, unsuspecting Smithsonian officials initially approved despite a house rule against showing political or religious material. That approval implied Smithsonian cosponsorship, which generated considerable heat from the scientific community.

The Smithsonian quickly withdrew its cosponsorship, stating: "We have determined that the content of the film is not consistent with the mission of the Smithsonian Institution's scientific research."[7] They allowed the film to be shown but turned down the payment.

THE UNCONGENIAL UNIVERSE 141

HOW COMMON IS LIFE IN THE UNIVERSE?

Let us take a look at the scientific facts about life in the universe, hopefully unbiased by theological considerations. Unfortunately, we only have one data point—Earth. Life has yet to be found anywhere but on Earth. Over a hundred planets beyond our solar system have been identified, with more being found regularly. None, so far, are likely to be suitable for complex life as we know it and certainly not for human life. This failure may be simply a matter of inadequate detector technology. However, the very fact that the powerful instruments of modern science, which can peer inside nuclei and out to the edge of the visible universe, have yet to find life outside Earth is already strong testimony that the galactic space around Earth is not exactly teeming with life.

Perhaps life may someday be confirmed on Mars or elsewhere in the solar system, such as under the ice on Jupiter's moon Europa or on Saturn's moon Titan. But such life undoubtedly will be at best primitive. Certainly humans cannot live on Mars or in an ocean on Europa without extensive life support. In fact, we very probably cannot live on more than the tiniest fraction of the planets in the universe. Not only are earthlike planets likely to be very rare; so are sunlike stars.

One often hears that our sun is a "typical star." This is wrong. In fact, 95 percent of all stars are less massive than the sun. Stars much more massive than the sun have short lifetimes. If life is to be widespread in the universe, then it will have to exist under a far wider range of conditions than exist on Earth. And, how likely then is the possibility of intelligent life?

The observations mentioned above imply that on the order of ten billion stars in the Milky Way may have planetary systems. While some form of life might have evolved in a large fraction of these systems, the very reasons that Gonzalez and Richards give for Earth being "privileged" make it very unlikely that humans could survive without extensive life support, even on those planets that might otherwise be suitable for some kind of life.

In recent years a new scientific discipline called *astrobiology* has appeared on the scene to study the possibilities of extraterrestrial life. This has brought together not only astronomers and biologists but philosophers and theologians to debate such issues as the definition of life and the impact the discovery of life elsewhere would have on human thinking.

Sufficient data to settle the question of life elsewhere are still lacking. As mentioned, a whole spectrum of views can be found among those working in this field. At one extreme we have what is called the *rare-earth position*, as exemplified by the book of that title by paleontologist Peter D. Ward and astronomer Donald Brownlee, published in 2000,[8] and *The Privileged Planet*, which was discussed above. In this view, complex forms of life are uncommon if not exceedingly rare in the universe.

The other end of the spectrum maintains the viewpoint that complex life could in fact be quite common. Astronomer David Darling summarized both positions in *Life Everywhere*, which appeared in 2001.[9] He argues that the rare-earth position is far too conservative, given what we currently know and don't know.

Both extremes and those in between agree that simple, primitive forms of life are likely to exist on an appreciable fraction of other planets. This conclusion is warranted by the discovery in recent years of new (but still DNA-based) forms of life on Earth, thriving under the most extreme conditions in deep-ocean vents, bubbling volcanic mud pots, frigid waters, and complete darkness. Indeed, life on Earth may have even begun under those conditions.

The real controversy is over the likelihood of complex, multicellular life. While microbial life is found over a wide range of conditions on Earth, the complex structures that make up animals and plants are very sensitive to their environments. Since we are hardly going to settle the issue here, let us just look at the flavor of the debate.

form the basis for life. Using anthropic arguments, astronomer Fred Hoyle predicted this energy level before it was confirmed experimentally.[18]

All these statements can be expressed in a unit-free way.

HOW SIGNIFICANT IS THE FINE-TUNING?

Let us take a look at these parameters to see how significant is the fine-tuning. The strength of the electromagnetic force is determined by a dimensionless parameter α called the *fine structure constant*, which depends on the value of the unit electric charge, that is, the magnitude of the charge of an electron conventionally designated by e.[19] The claim is that α has been fine-tuned far from its natural value in order that we have stars sufficiently long-lived for life to evolve (item 1 above).

However, α is not a constant. We now know from the highly successful standard model of particles and forces that α and the strengths of the other elementary forces vary with energy and must have changed very rapidly during the first moments of the big bang when the temperature changed by many orders of magnitude in a tiny fraction of a second. According to current understanding, in the very high-temperature environment at the beginning of the big bang, the four known forces were unified as one force. As was discussed in the previous chapter, the universe can be reasonably assumed to have started in a state of perfect symmetry, the symmetry of the "nothing" from which it arose. So, α began with its natural value; in particular, gravity and electromagnetism were of equal strength. That symmetry, however, was unstable and, as the universe cooled, a process called *spontaneous symmetry breaking* resulted in the forces separating into the four basic kinds we experience at much lower energies today, and their strengths evolved to their current values. They were not fine-tuned. Stellar formation and, thus, life had to simply wait for the forces to separate sufficiently. That wait was actually a tiny fraction of a second.

The forces continued to separate as the universe continued to cool, but this was so slow that for all practical purposes on a human timescale, the strengths of the various forces can be regarded as constant.

Only four parameters are needed to specify the broad features of the universe as it exists today: the masses of the electron and proton and the current strengths of the electromagnetic and strong interactions.[20] (The strength of gravity enters through the proton mass, by convention.) I have studied how the minimum lifetime of a typical star depends on the first three of these parameters.[21] Varying them randomly in a range of ten orders of magnitude around their present values, I find that over half of the stars will have lifetimes exceeding a billion years. Large stars need to live tens of millions of years or more to allow for the fabrication of heavy elements. Smaller stars, such as our sun, also need about a billion years to allow life to develop within their solar system of planets. Earth did not even form until nine billion years after the big bang. The requirement of long-lived stars is easily met for a wide range of possible parameters. The universe is certainly not fine-tuned for this characteristic.

One of the many major flaws with most studies of the anthropic coincidences is that the investigators vary a single parameter while assuming all the others remain fixed. They further compound this mistake by proceeding to calculate meaningless probabilities based on the grossly erroneous assumption that all the parameters are independent.[22] In my study I took care to allow all the parameters to vary at the same time.

Physicist Anthony Aguire has independently examined the universes that result when six cosmological parameters are simultaneously varied by orders of magnitude, and found he could construct cosmologies in which "stars, planets, and intelligent life can plausibly arise."[23] Physicist Craig Hogan has done another independent analysis that leads to similar conclusions.[24] And, theoretical physicists at Kyoto University in Japan have shown that heavy elements needed for life will be present in even the earliest stars independent of what the exact parameters for star formation may have been.[25]

The current standard model of elementary particles and forces contains about twenty-four parameters that currently are not determined by the theory but must be inferred from experiments. This is not as bad as it might seem, since the model accurately describes thousands of data points. In any case, only four parameters are needed to specify most properties of matter. These are the masses of the electron and the two quarks ("up" and "down") that constitute protons and neutrons, and a universal strength parameter from which the value α and the other force strengths are obtained. Ultimately, it is hoped that all the basic parameters will be determined by theories that unify gravity with the standard model, for example, *string theory*.[26] We must wait to see if the calculated masses of the electron and neutron come out to satisfy coincidences 3 and 4 above.

ARE CARBON AND ORGANIC MOLECULES FINE-TUNED?

Let us next take a more detailed look at coincidence 5, which asserts that fine-tuning is needed to produce carbon, the primary building block of life. Astronomer Fred Hoyle used anthropic arguments to successfully predict the presence of a nuclear energy level in carbon at 7.65 million electron-volts. However, M. Livio and collaborators have shown that the production of carbon in stars does not depend sensitively on that nuclear energy level. Rather it hinges on the radioactive state of a carbon nucleus formed out of three helium nuclei, which misses being too high for carbon production by only 20 percent.[27] Nobel laureate physicist Steven Weinberg has noted that this "is not such a close call after all."[28]

The chemical elements carbon and oxygen are among the easiest to produce in the nuclear reactions that take place in dying stars. The main energy source in stars is the fusion of hydrogen into helium. The helium nucleus, composed of two protons and two neutrons and symbolized by $_2He^4$, is highly stable—as predicted by the rules of quantum mechanics.[29] Two helium nuclei can fuse to give a beryllium nucleus,

$$_2He^4 + _2He^4 \rightarrow _4Be^8$$

Another helium then can fuse with the beryllium to produce carbon,

$$_2He^4 + _4Be^8 \rightarrow _6C^{12}$$

And yet another helium can fuse with the carbon to give oxygen,

$$_2He^4 + _6C^{12} \rightarrow _8O^{16}$$

Each of these product nuclei is also very stable and so will survive indefinitely. When the star finally runs out of energy these elements among others in the periodic table, especially iron, are distributed into the space between stars, either by evaporation or, in the case of very massive stars, enormous explosions called supernovae.[30]

In short, no fine-tuning is necessary for the production of carbon, oxygen, and the other basic elements of life. They are in fact the elements that are among the easiest to form by common nuclear reactions.

So, too, are the molecular ingredients of life easy to produce. In a remarkably simple experiment in 1952, which took only weeks to assemble, graduate student Stanley Miller, working under the renowned chemist Harold Urey, sent a 60,000-volt electrical spark, simulating lightning, through a flask containing a gas of methane, ammonia, hydrogen, and water vapor. At the time, this was thought to simulate the atmosphere of early Earth. The by-product contained amino acids, the basic chemical sub-unit of proteins, and other raw materials of life.[31]

We now know that Miller's gas mixture did not accurately represent the Earth's atmosphere at the likely time that life originated. Some theists have seized on this to dismiss the importance of the experiment.[32] But, they miss the point, which is that the complex, carbon-based molecules that occur in living matter can be readily produced by chemical reactions involving simpler substances. This is another example of how simplicity can beget complexity, contrary to the claims of creationists.

Astrobiologists have now demonstrated that organic molecules occur under a wide range of conditions, including those that existed on the early Earth and those existing in space. Space origins are confirmed by the observation of these molecules in meteorites analyzed immediately after striking Earth so that effects of contamination by earthly matter are minimal. Perhaps the first ingredients of life came from space after Earth formed.[33]

IS THE VACUUM ENERGY FINE-TUNED?

Next, let us examine the claim that the vacuum energy of the universe is fine-tuned. Normally we think of the vacuum as being empty of matter and energy. However, according to general relativity, gravitational energy can be stored in the curvature of empty space. Furthermore, quantum mechanics implies that a vacuum could contain a minimum *zero-point energy*.

Weinberg referred to this as the *cosmological constant problem*, since any vacuum energy density is equivalent to the parameter in Einstein's theory of general relativity called the cosmological constant that relates to the curvature of empty space-time.[34] A better term is *vacuum energy problem.*

Crude calculations gave a value for the vacuum energy density that is some 120 orders of magnitude greater than its maximum value from observations. Since this density is constant, it would seem to have been fine-tuned with this precision from the early universe, so that its value today allowed for the existence of life.

Until recent years, it was thought that the cosmological constant is exactly zero, in which case there was no need for fine-tuning, although no theoretical reason was known. However, in 1998, two independent research groups studying distant supernovae were astonished to discover that the current expansion of the universe is *accelerating.*[35] More recent observations from other investigators have confirmed this result. The universe is falling up! The source of this cosmic acceleration may be some still-unidentified *dark energy*, which constitutes 70 percent of the mass of the universe. One possible mechanism is gravitational repulsion by means of the cosmological constant, that is, by way of a vacuum energy field, which is allowed by general relativity.

If that is the case, then the cosmological constant problem resurfaces. In the meantime, we now have plausible reasons to suspect that the original calculation was incomplete and that a proper calculation will give zero for the vacuum energy density.[36] Until these newer estimates are shown to be wrong, we cannot conclude that the vacuum is fine-tuned for life and we have no particularly strong need to invoke a designer deity.

But, then, what is responsible for cosmic acceleration, that is, what is the nature of the dark energy? A cosmological constant is not the only possible source of gravitational repulsion. According to general relativity, any matter field will be repulsive if its pressure is sufficiently negative. Theorists have proposed that the dark energy may be a matter field, called *quintessence*, which requires no fine-tuning.[37] Finally, it should be noted that cosmologists are still not totally convinced that dark energy must be invoked to account for the observed cosmic acceleration and have proposed alternate mechanisms.

OTHER FORMS OF LIFE?

Carbon would seem to be the chemical element best suited to act as the building block for the type of complex molecular systems that develop lifelike qualities. Even today, new materials assembled from carbon atoms exhibit remarkable, unexpected properties, from superconductivity to ferromagnetism. We expect any life found in our universe to be carbon-based, or at least based on heavy element chemistry.

But that need not be true in every conceivable universe. Even if all the forms of life discovered in our universe turn out to be of the same basic structure, it does not follow that life is impossible under any other arrangement of physical laws and constants. According to the scenario I mentioned briefly in the last chapter, certain laws of physics are likely to be common to any universe born out of empty space-time, but others along with many physical constants may be the result of a random process called spontaneous symmetry breaking.

The possibility of other laws and constants is fatal to the fine-tuning argument. Philosopher Gilbert Fulmer has shown that the fine-tuning argument is logically incoherent.[38] Simplifying his more detailed analysis, the fine-tuning argument requires that the set of facts for our universe, {U1}, could have been a different set, {U2}. But in that case, we cannot use {U1}, which is all we know, to say anything about {U2}. (See epigraph to this chapter by David Hume.)

We can only speculate what form life might take on another planet, with different conditions. It would be wonderful to have more examples of life, but we do not. And, any speculation about what form life might take in a universe with a different electron mass, electromagnetic interaction strength, or different laws of physics is even more problematical. We simply do not have the knowledge to say whether life of *some* sort would not occur under different circumstances.

Theists who argue that the universe is fine-tuned to earthly life have the burden of proving that no other form of life is possible, not just on other planets in our universe but in every conceivable universe that has different physical parameters. They have provided no such proof and it would seem that such a proof is impossible.

In fact, the whole argument from fine-tuning ultimately makes no sense. As my friend Martin Wagner notes, all physical parameters are irrelevant to an omnipotent God. "He could have created us to live in hard vacuum if he wanted."[39]

WASTE

The anthropic argument for the existence of God can be turned on its head to provide an argument against the existence of God. If God created a universe with at least one major purpose being the development of *human* life, then it is reasonable to expect that the universe should be congenial to *human* life. Now, you might say that God may have had other purposes besides humanity. As has been noted several times in this book, apologists can always invent a god who is consistent with the data. One certainly can imagine a god for whom humanity is not very high on the agenda and who put us off in a minuscule, obscure corner of the universe. However, this is not the God of Judaism, Christianity, and Islam, who places great value on the human being and supposedly created us in his image. Why would God send his only son to die an agonizing death to redeem an insignificant bit of carbon?

If the universe were congenial to human life, then you would expect it to be easy for humanlike life to develop and survive throughout the universe.

nation would contract, as measured from their own reference frame. An astronomer on Earth would measure the usual distance between astronomical objects but would observe the spaceship clocks to slow down and the astronauts to age more slowly.

Suppose we were able to build a spaceship that could accelerate at a constant one g, that is, at the acceleration of gravity on Earth, which would also nicely provide artificial gravity for the astronauts. That ship would reach Alpha Centauri in five years' Earth-time while only a bit over two years would elapse in ship-time. In eleven years' ship-time it could reach the center of our galaxy. But during that time, almost 27,000 years would have passed on Earth. In fifteen years' ship-time the astronauts could reach Andromeda, 2.4 million light-years away. By then, since most of the trip was at near the speed of light relative to Earth, 2.4 million years would have gone by back on Earth. After experiencing the passage of twenty-three years, the astronauts would actually pass the edge of the universe currently observable from Earth, but 13.7 billion years would have elapsed in the reference frame of a long-dead Earth.

If the astronauts wish to stop at any of these places to explore for earthlike planets, then the times must be doubled, since they could accelerate only during the first half of the trip and then would have to decelerate for the second half.

The unavoidable fact seems to be that any humans exploring the universe will effectively cut themselves off from Earth. Even if they traveled to the center of the Milky Way and back, aging forty-four years in the process, they would return to an Earth 104,000 years in the future as measured on Earth clocks. Basically, any humans traveling to the stars would forever leave behind their families, their society, and even their species.

Notice that I have not asserted any technological limitations to argue that spaceflight to distant stars and galaxies is impossible. While a method for accelerating a spaceship to near the speed of light is beyond any technology we can currently imagine, we cannot rule that out for future generations. Authors also speculate about traveling through *wormholes*, tunnels through space-time that act as shortcuts to other parts of the universe.[40] I don't know if that will ever prove possible, but I doubt it.

But, suppose such explorations do someday take place. How earthlike must a planet be for humans to be able to live there? Life on Earth evolved under the very special set of conditions that exist here. We are adapted to live on Earth and not just anywhere in space. We would not be overly pessimistic in guessing that space travelers would have to travel tens of thousands of light-years, at the minimum, before finding a planet they could live on without massive life support.

The suggestion is frequently made that humanity might someday live in outer space, inside space stations orbiting Earth and other planets. However, even if these space stations duplicate all the conditions on Earth, they may not be able to deal with the cosmic rays from which we on Earth are shielded by the atmosphere. The same threat would seem to prohibit lengthy space travel of the type described earlier. Even the Mars missions people dream about would very possibly expose astronauts to life-shortening radiation poisoning. Traveling outside the solar system would kill them.

Perhaps future technologies will solve this problem, too. Maybe genetic engineering will make new kinds of humans, really a new species, suitable for space travel. And, of course, we can always send automatons.

Whatever the imagined possibilities, the strong conclusion is that humans are not constructed to live anywhere but on this tiny blue speck in a vast universe. Maybe many similar specks exist throughout the universe, but Homo sapiens is unlikely to ever find them. Our species is probably marooned in space, on space-ship Earth, and likely to go extinct long before the sun burns its last hydrogen atom.

However, once we give up the idea that we are special children of God, we can see ourselves as a link in the chain of evolution.

Our descendants, genetically engineered or made of titanium and silicon, unhampered by our brief life spans, may reach other planets. And if we do it right they hopefully will be smarter, kinder, more rational, and free of the superstitions that plague us and threaten our very survival even for a few more centuries.

Even taking the most optimistic view of the future of humankind, though, it is hard to conclude that the universe was created with a special, cosmic purpose for humanity. It seems inconceivable that a creator exists who has a special love for humanity, and then just relegated it to a tiny point in space and time. The data strongly suggest otherwise. Indeed, the universe looks very much like it was produced with no attention whatsoever paid to humanity.

When we take even the most optimistic estimates of the density of intelligent life of all kinds in the universe, those civilizations are still separated by enormous distances with nothing but wasted space in between. It is also hard to believe that the universe was created with a special, cosmic purpose for intelligent life of any kind.

A LIFE PRINCIPLE?

Despite the apparent uncongeniality of the universe to complex life, life is present and some people still insist that this alone is remarkable. Physicist Paul Davies suggests that perhaps a *life principle* is "written into the laws of physics" or "built into the nature of the universe."

But nowhere in current physics, chemistry, or biology do we find any sign of a fundamental life principle, some *élan vital* that distinguishes life from nonlife. Davies speculates, "A felicitous mix of law and chance might be generalised to cosmology, producing directional evolution from simple states, through complex, to life and mind."[41]

Davies shares this notion with biologists Christian de Duve[42] and Stuart Kauffman.[43] These authors all seem to view the life principle as some previously unrecognized, holistic, teleological law of nature, although what this may be is not at all clear from their highly speculative writings. As discussed in chapter 3, Nancey Murphy and other theologians admit that the traditional notion of a separate soul and body is no longer viable given the evidence from neuroscience. But, being theologians they have to find God somewhere. If they conclude God does not exist they are quickly out of a job. Some have put their stock in what they call "nonreductive physicalism." They think they can find a place for God and the soul therein.[44]

However, any life principle, if it exists, may be one of the type of so-called emergent principles found in chaos and complexity theory that naturally arise from the nonlinear, dissipative, but still purely local interactions of material particles.[45] These cannot be called new laws of physics since they follow from already existing laws, if not by direct, mathematical proof, then by computer simulations that involve no new principles. Indeed, as we have seen, such simulations indicate that complexity evolves from simplicity by familiar, purely reductive physical processes without the aid of any overarching holistic guiding principle.[46]

A TINY POCKET OF COMPLEXITY

It is commonly thought that the universe is an intricately complex place. However, taking an overview we can see that this is a selection effect resulting from the fact that we and our planet are relatively complex. Most of the matter and energy of the universe exhibits little structure and shows no sign of design. We noted above that 96 percent of the mass of the universe appears to be composed of dark matter and dark energy whose exact natures are unknown but that are definitely not composed of familiar atomic matter. As far as we can tell, these components have little structure.

The very low-energy photons in the cosmic microwave background radiation are a billion times more plentiful than the atoms in galaxies. These particles are spread uniformly throughout the universe to one part in a hundred thousand. They move around almost completely randomly, as if they were a gas in thermal equilibrium having maximum entropy and at a temperature only three degrees above absolute zero on the Kelvin scale. The little structure that is seen is understood as the remnant of random fluctuations that took place in the early universe and helped trigger galaxy formation. Again, absence of design is evident.

Physicist Max Tegmark has argued that the universe contains almost no information, that is, it has on the whole no structure.[47] He suggests that the large information content that we humans perceive results from our subjective viewpoint. According to quantum mechanics, the universe is perfectly random, a superposition of all possible realities. However, the very act of observation selects out only one of those realities. Some quantum mystics, such as the popular author Deepak Chopra, interpret this as an ability of humans to "make our own reality."[48] However, the evidence clearly indicates otherwise.[49] If we could make our own reality we would all continue to look like we did when we were twenty. But even Chopra is aging along with the rest of us. The reality that is selected by our observations is just a toss of the dice.

Even if Tegmark is off the mark, any huge, random universe, regardless of its properties, will naturally develop at least a few tiny pockets of complexity within a vast sea of chaos, which is just what we seem to see in our universe. We do not need either a designer or multiple universes to account for such rare deviations as are consistent with chance.

It is rather amusing that theists make two contradictory arguments for life requiring a creator. Sometimes you hear these from the same people. In the fine-tuning argument, the universe is so *congenial* to life that the universe must have been created with life in mind. But, if it is so congenial, then we should expect life to evolve by natural processes and a sustaining God is unnecessary. In the second argument, the universe is so *uncongenial* to life that life could not have occurred by natural processes and so must have been created and be sustained by the constant actions of God. There is a third and much simpler possibility that fits the data far better; we are just the product of circumstance and chance.

If God created matter with human life in mind, he did not use very much of it for his purpose. If God created order, he did not make much of that either. The observed universe and the laws and parameters of physics look just as they can be expected to look if there is no God. From this we can conclude, beyond a reasonable doubt, that such a God does not exist.

In a paper appearing just as this book was going to press, Roni Harnik, Graham Kribs, and Gilad Perez have constructed a universe without any weak nuclear interactions.[50] They find that this universe undergoes big bang nucleosynthesis, matter domination, structure formation, and star formation. Stars burn for billions of years, synthesizing elements up to iron and undergoing supernova explosions, dispersing heavy elements into the interstellar medium. Chemistry and nuclear physics are essentially unchanged. This is one more example, to be added to those discussed above, where a claim that certain parameters of the universe, in this case those of the weak interaction, are fine-tuned for life.

NOTES

1. Hugh Ross, *The Creator and the Cosmos: How the Greatest Scientific Discoveries of the Century Reveal God*, rev. ed. (Colorado Springs: Navpress, 1995), pp. 138–45.

2. Guillermo Gonzalez and Jay W. Richards, *The Privileged Planet: How Our Place in the Cosmos Is Designed for Discovery* (Washington, DC: Regnery, 2004), p. 335.

3. Clifford M. Will, *Was Einstein Right? Putting General Relativity to the Test* (New York: Basic Books, 1986).

4. Voltaire, *Candide* (1759), as quoted in the *TalkOrigins* archive, online at http://www.talkorigins.org/indexcc/CI/CI302.html (accessed June 5, 2005). The full text in English is available online at http://www.literature.org/authors/voltaire/candide/ (accessed June 5, 2005).

5. Center for Science and Culture, online at http://www.discovery.org/csc/fellows.php (accessed June 6, 2005).

6. Barbara Forrest and Paul R. Gross, *Creationism's Trojan Horse: The Wedge of Intelligent Design* (Oxford and New York: Oxford University Press, 2004).

7. Tommy Nguyen, "Smithsonian Distances Itself from Controversial Film," *Washington Post*, June 2, 2005.

8. Peter D. Ward and Donald Brownlee, *Rare Earth: Why Complex Life Is Uncommon in the Universe* (New York: Copernicus, 2000).

9. David J. Darling, *Life Everywhere: The Maverick Science of Astrobiology* (New York: Basic Books, 2001).

10. Ibid., pp. 95–110.

11. Richard Swinburne, "Argument from the Fine-Tuning of the Universe" in *Modern Cosmology and Philosophy*, ed. John Leslie (Amherst, NY: Prometheus Books, 1998), pp. 160–79; George Ellis, *Before the Beginning: Cosmology Explained* (London, New York: Boyars/Bowerdean, 1993); Ross, *The Creator and the Cosmos*; Patrick Glynn, *God: The Evidence* (Rocklin, CA: Prima Publishing, 1997); Dean L. Overman, *A Case Against Accident and Self-Organization* (New York, Oxford: Rowman & Littlefield, 1997).

12. Brandon Carter, "Large Number Coincidences and the Anthropic Principle in Cosmology" in *Confrontation of Cosmological Theory with Astronomical Data*, ed. M. S. Longair (Dordrecht: Reidel, 1974), pp. 291–98, reprinted in *Modern Cosmology and Philosophy*, ed. John Leslie (Amherst, NY: Prometheus Books, 1998), pp. 131–39.

13. John D. Barrow and Frank J. Tipler, *The Anthropic Cosmological Principle* (Oxford: Oxford University Press, 1986).

14. See "Anthropics" online at http://www.colorado.edu/philosophy/vstenger/anthro.html (accessed June 10, 2005).

15. J. J. Davis, "The Design Argument, Cosmic 'Fine Tuning,' and the Anthropic Principle," *Philosophy of Religion* 22 (1987): 139–50.

16. Neil A. Manson, "There Is No Adequate Definition of 'Fine-tuned for Life,'" *Inquiry* 43 (2000): 341–52.

17. Robert Klee, "The Revenge of Pythagoras: How a Mathematical Sharp Practice Undermines the Contemporary Design Argument in Astrophysical Cosmology," *British Journal for the Philosophy of Science* 53 (2002): 331–54.

18. F. Hoyle et al., "A State in C12 Predicted from Astrophysical Evidence," *Physical Review Letters* 92 (1953): 1095.

19. $\alpha = e^2/\hbar c$ or $e^2/4\pi\varepsilon_o\hbar c$, depending on unit system, where $\hbar = h/2\pi$, h is Planck's constant, c is the speed of light, and ε_o is an electrical constant called the permittivity of free space. The low-energy value of α is $1/137$.

20. W. H. Press and A. P. Lightman, "Dependence of Macrophysical Phenomena on the Values of the Fundamental Constants," *Philosophical Transactions of the Royal Society of London* A 310 (1983): 323–36; B. J. Carr and M. J. Rees, "The Anthropic Principle and the Structure of the Physical World," *Nature* 278 (1979): 606–12.

21. Victor J. Stenger, *The Unconscious Quantum: Metaphysics in Modern Physics and Cosmology* (Amherst, NY: Prometheus Books, 1995), pp. 236–38; "Natural Explanations for the Anthropic Coincidences," *Philo* 3, no. 2 (2001): 50–67.

22. See, for example, R. Totten, "The Intelligent Design of the Cosmos: A Mathematical Proof" (2000), http://www.geocities.com/worldview_3/mathprfcosmos.html (accessed February 6, 2005).

23. Anthony Aguire, "The Cold Big-Bang Cosmology as a Counter-example to Several Anthropic Arguments," *Physical Review* D64 (2001): 083508.

24. Craig J. Hogan, "Why the Universe Is Just So," *Reviews of Modern Physics* 72 (2000): 1149–61.

25. Takashi Nakamura, H. Uehara, and T. Chiba, "The Minimum Mass of the First Stars and the Anthropic Principle," *Progress of Theoretical Physics* 97 (1997): 169–71.

26. Gordon L. Kane, Michael J. Perry, and Anna N. Zytkow, "The Beginning of the End of the Anthropic Principle," *New Astronomy* 7 (2002): 45–53.

27. M. Livio et al., "The Anthropic Significance of the Existence of an Excited State of ^{12}C," *Nature* 340 (1989): 281–84.

28. Steven Weinberg, "A Designer Universe?" *New York Review of Books*, October 21, 1999. Reprinted in the *Skeptical Inquirer* (September/October 2001): 64–68.

29. The subscript indicates the number of protons, the superscript the number of protons and neutrons. The total of each number is conserved in a nuclear reaction, as can be seen in reactions discussed in the text.

30. Elements beyond iron are only produced in the massive stars that produce supernovae.

31. Stanley L. Miller, "A Production of Amino Acids under Possible Primitive Earth Conditions," *Science* 117 (1953): 528–29.

32. Overman, *A Case Against Accident and Self-Organization*, pp. 41–49.

33. Darling, *Life Everywhere*, pp. 33–51, and references therein.

34. Steven Weinberg, "The Cosmological Constant Problem," *Reviews of Modern Physics* 61 (1989): 1–23.

35. A. Reiss et al., "Observational Evidence from Supernovae for an Accelerating Universe and a Cosmological Constant," *Astronomical Journal* 116 (1998): 1009–38; S. Perlmutter et al., "Measurements of Omega and Lambda from 42 High-Redshift Supernovae," *Astrophysical Journal* 517 (1999): 565–86.

36. Victor J. Stenger, *The Comprehensible Cosmos: Where Do the Laws of Physics Come From?* (Amherst, NY: Prometheus Books, 2006).

37. Lawrence Krauss, *Quintessence: The Mystery of the Missing Mass in the Universe* (New York: Basic Books, 2000).

38. Gilbert Fulmer, "A Fatal Logical Flaw in Anthropic Design Principle Arguments," *International Journal for Philosophy of Religion* 49 (2001): 101–10.

39. Martin Wagner, private communication.

40. Kip S. Thorne, *Black Holes & Time Warps: Einstein's Outrageous Legacy* (New York: Norton, 1994).

41. Paul Davies, *The Cosmic Blueprint* (New York: Simon and Schuster, 1988; Radnor, PA: Templeton Foundation Press, 2004); "Multiverse or Design: Reflections on a Third Way," *Proceedings of Universe or Multiverse?* Stanford University (March 2003), http://aca.mq.edu.au/PaulDavies/Multiverse_StanfordUniv_March2003.pdf (accessed January 4, 2005).

42. Christian de Duve, *Vital Dust* (New York: Basic Books, 1995).

43. Stuart Kauffman, *At Home in the Universe: The Search for the Laws of Self-Organization and Complexity* (New York and Oxford: Oxford University Press, 1995).

44. Warren S. Brown, Nancey Murphy, and H. Newton Malony, eds., *Whatever Happened to the Soul? Scientific and Theological Portraits of Human Nature* (Minneapolis: Fortress Press, 1998).

45. Steven Johnson, *Emergence: The Connected Lives of Ants, Brains, Cities, and Software* (New York: Touchstone, 2001).

46. Christoph Adami, *Introduction to Artificial Life* (New York: Springer, 1998); Christoph Adami, Charles Ofria, and Travis C. Collier, "Evolution of Biological Complexity," *Proceedings of the National Academy of Sciences USA* 97 (2000): 4463–68.

47. Max Tegmark, "Does the Universe In Fact Contain Almost No Information?" *Foundations of Physics Letters* 9, no. 1 (1996): 25–42.

48. Deepak Chopra, *Quantum Healing: Exploring the Frontiers of Mind/Body Medicine* (New York: Bantam, 1989); *Ageless Body, Timeless Mind: The Quantum Alternative to Growing Old* (New York: Random House, 1993).

49. Stenger, *The Unconscious Quantum*.

50. Roni Harnik, Graham D. Kribs, and Gilad Perez, "A Universe without Weak Interactions," *Physical Review* D74 (2006): 035006.

Chapter 6

THE FAILURES OF REVELATION

If the statements it [the Bible] contains concerning matters of history and science can be proven by extrabiblical records, by ancient documents recovered through archaeological digs, or by the established facts of modern science to be contrary to the truth, then there is grave doubt as to its trustworthiness in matters of religion. In other words, if the biblical record can be proved fallible in areas of fact that can be verified, then it is hardly to be trusted in areas where it cannot be tested.
—Archer L. Gleason[1]

TESTING REVELATION

The God of Jews, Christians, and Muslims is believed to communicate with humanity. Mystics of all faiths and in all ages have reported such communication. The knowledge they claim to have received from God fills religious literature. While much of the material is esoteric and not readily confirmable, we can reasonably expect that some revealed wisdom should be amenable to empirical verification. This is especially true for statements about the observable world and physical events. We should be able to find remarkable examples where specific information about the world, which was unknown to science at the time of the revelation, would later be confirmed by observation. We should also be able to find numerous cases of successful predictions of future events that have no plausible alternate explanation.

Instead we find the opposite. Scriptures and other records of claimed revelations contain many disagreements with science about the physical world. These are not just disagreements about "theories," such as biological evolution as covered in chapter 2, but disagreements with now well-established empirical facts. (Well, evolution is an established empirical fact, too, but this has not stopped it from being politically controversial.)

Similarly, the records of claimed revelations contain no prediction of a future event that cannot be plausibly accounted for without recourse to the supernatural.

We will discuss three types of failures of revelation. In the first, we will see that no information supposedly gained during a mystical or religious experience, which could not have been otherwise known to the individual claiming the experience, has ever been confirmed. In the second type of revelation failure, the scriptures will be seen to contain gross errors of scientific fact. Third, we will see that not a single risky biblical prophecy can be shown, by objective means, to have been fulfilled. Finally we will show that lack of physical evidence proves conclusively that important biblical tales, such as the Exodus and the events surrounding Jesus' birth and death, cannot have occurred on the scale and manner described in the Bible. From all of this, it follows that the scriptures and reported religious experiences are not sources of revealed information.

Now, once again, standard scientific criteria are being applied in drawing these conclusions—the same criteria that are used to test all extraordinary claims. Personal testimonials and anecdotal stories have little or no value as evidence for the truth of extraordinary claims. Poorly controlled experiments are similarly useless. Furthermore, predictions of future events have little or no value unless those predictions are risky, that is, they could have turned out otherwise. Predicting the sun will come up tomorrow is not risky. Predicting it will not—now that's risky! And, although this may seem an obvious requirement, prediction must be made before the fact. Many of the claimed fulfilled prophecies in scriptures were actually made after the prophesized events took place.

RELIGIOUS EXPERIENCES

One place where truly spiritual revelation would be expected to produce testable consequences is with so-called religious experiences. Throughout history, people have claimed deep, life-altering mystical experiences and formed prophecies based on their visions. They say that they have been in touch with God or some other form of higher reality. I am convinced that many are sincere in that belief (television evangelists excepted). However, without independent confirmation, the reported experiences could have been all in their heads.

As was the case for the claimed special powers of the mind discussed in chapter 3, ways can be conceived to test for supernatural involvement in a religious experience. Once again, despite the widespread belief that science cannot deal with spiritual phenomena, it is really very simple. If a person undergoes a religious experience that truly places her in communication with some reality from beyond the material world, then we may reasonably expect that person to have gained some deep, new knowledge about the world that can be checked against the empirical facts.

Now, typically, the person having a religious experience returns with messages from beyond about how we humans should all love and care for one another, be kind to animals, preserve the environment, and not eat too much red meat. As seen in chapter 3, purely material brain processes can produce the same experiences as reported in a mystical experience. Indeed, such experiences can be induced by various physical and chemical means. In short, the mere occurrence of a religious experience is no evidence for a supernatural event.

Suppose, however, that instead of simple homilies someone undergoing an epiphany gains new knowledge that she could not have possibly obtained by purely physical means. For example, imagine that someone in the twentieth century had a vision that foresaw that on December 26, 2004, a tsunami in the Indian Ocean would kill tens of thousands of people. If that had happened, we would take seriously the notion that some power beyond the material world does indeed exist. In short, the validity of an otherworldly component to a religious experience is readily verifiable.

Despite many stories, however, no such report has stood up under scientific scrutiny. The prophecies of mystics have been either too vague to constitute a reasonable test, or downright failures. Just consider how many times throughout history that the end of the world has been proclaimed, with specific dates usually given. The world is still here.

Reported religious experiences are wholly unremarkable despite the cosmic proportions of the claim. We saw in chapter 3 that no successful (meaning statistically significant in ruling out all ordinary explanations) empirical tests for extrasensory perception, mind-over-matter, the efficacy of prayer, and other mystical or semimystical claims can be found in reputable scientific literature. Similarly, special revelation through religious experiences has not become part of common scientific knowledge.

It does not suffice to say that perhaps these phenomena may

still exist at some low level that has not yet been detected, or that the issue is still controversial. Believers can accuse nonbelievers of being dogmatically skeptical and unwilling to "open their eyes to the truth." But our eyes are open and we see no convincing evidence for phenomena that under the God hypothesis would be expected to hit us all square in the face. If the religious experience were as deeply significant as the monotheistic religions have taught, then data would exist that even the most die-hard skeptic could not ignore.

Now, it might be argued that God has not chosen to reveal physical facts about nature that can be tested empirically. But surely the God of the monotheisms is believed to reveal moral knowledge. And, that moral knowledge, as we will discuss in chapter 7, is empirically testable. Indeed we will find that the hypothesis of a God who provides moral knowledge is falsified by the observable fact that many of the moral teachings found in scriptures that are supposed revelations are not obeyed by even the most pious faithful.

A God who provides humans with important knowledge that they cannot obtain by material means should have produced testable evidence for his existence by now. He has not. The evidence points to the opposite conclusion. We can say with some confidence that such a God does not exist.

SCRIPTURE AND SCIENCE

The next places we will look for scientific evidence of revelation are in the scriptures. The least ambiguous and most egregious scientific errors in these books can be found in their references to phenomena now studied in the scientific fields of astronomy, cosmology, and biology.

One often hears the claim that *big bang cosmology* confirms what is written in Genesis, thus "proving" the existence of the

God of the Bible. However, almost all cultures and religions have their creation myths, and we need to compare these, as well as the scientific facts, with the details presented in the Bible.

With thousands of religions, past and present, we cannot possibly list every creation story. So let us select just a few, which should at least illustrate that the Bible is not the sole source of creation narratives.

An ancient Chinese myth tells us that everything started in chaos. The universe was like a black egg (a black hole?). A god named Pan Gu, wielding an axe, breaks the egg and the heavens begin to expand. The fleas and lice on Pan Gu's body evolve into humankind.

In the Apache tale, nothing existed in the beginning—no Earth, no sky, no sun, no moon. Out of the darkness a thin disk appeared within which sat a bearded man, the Creator, the One Who Lives Above.

The Tahitian story begins with Taaroa, who just was. He found himself all alone in the void. He calls in every direction and nothing replies, so he changes himself into the universe.

In the Bible and Qur'an, a presumably preexisting God creates the universe in six days. Following the story in Genesis, Earth is created on the first day. Four days later, God creates the sun, moon, and stars.

Now, what does science tell us about the origin of the universe? In recent years, observational cosmology has grown into an astonishingly precise science. The totality of data from a range of telescopes and other instruments, on the ground and in space, now solidly support the so-called big bang model of an expanding universe. In that model, the visible matter found in tens of billions of giant galaxies and in much greater amounts of invisible "dark matter" and "dark energy" emerged from a tiny volume of space some 13.7 billion years ago by current astronomical estimates.

Observations indicate that Earth was not formed until nine

billion years after the initiation of the big bang, grossly contra-
dicting the sequence of events presented in Genesis. Furthermore,
the Bible seems to imply that the creation happened rather
recently—on the order of ten thousand years ago. At that time,
scripture says that all the "kinds" of living things were created and
since then have remained immutable, in disagreement with evolution. Throughout the Bible, the universe is referred to as a "firmament" that sits above a flat, immovable Earth.[2]

We see little resemblance in Genesis to the picture drawn by
contemporary science. All these facts can lead to only one conclusion: the biblical version of creation is dead wrong.[3]

The Chinese myth described above provides an account closer
to the scientific one than the Bible's myth, picturing an
expanding universe beginning in complete chaos and suggesting
the evolution of life. However, it can hardly be considered an
accurate description of the scientific data.

Theists often bring up the fact that a Catholic priest, Georges
Lemaître, first proposed the big bang in 1927. That's true; but
Lemaître was an eminent astronomer as well as a priest, and
while the notion of a divine creation was undoubtedly part of his
thinking, his proposal was based on good science rather than theology. As mentioned in chapter 4, he strongly advised the pope
not to make the big bang an infallible teaching of the Church.

Skeptical literature lists a range of other types of scientific
errors in scriptures, and places where pronouncements of
dubious scientific merit are made, such as pi having the value of
three. However, we have no need to discuss these, since biblical
language is vague and ambiguous. Apologists can always find
ways to make most biblical errors sound less damaging. Certainly
we might suppose that God, if he exists, speaks to people in the
language they understand. Ancient peoples cannot be expected to
have understood the language of modern science or have needed
an exact value of pi (except for the builders of great monuments
like the pyramids).

Still, the argument can be cast in the following terms, which by now should sound familiar: Our observations, in this case our reading of biblical and Qur'anic statements about the natural world, look exactly as you would expect them to look if there was no new knowledge being revealed—just what was the human understanding of the day. That is, they look as if there is no God who speaks to humanity through scriptures or other revelations.

THE JESUS PROPHECIES

It could have been different. The scriptures might have contained revelations that, while incomprehensible to people at the time of the revelation, may have still been recorded as mysterious, esoteric knowledge. That knowledge then might have become less esoteric as science and the other knowledge arts, such as history, developed higher levels of sophistication.

For example, suppose the New Testament somewhere contained the following passage: "Before two millennia shall pass since the birth of our Lord, a man will stand on another world within the firmament and he will smite a tiny orb with his staff such that it will fly from sight."[4] Obviously no mere mortal in Jesus' day could have anticipated that in two thousand years men would walk on the moon. Nor would he be expected to know anything about golf.

But, we have no risky prediction anywhere in the scriptures that has come true. Of course, preachers have disingenuously told their flocks that many biblical prophecies have been fulfilled.

In *Evidence That Demands a Verdict*,[5] written three decades ago, Josh McDowell of the Campus Crusade for Christ claimed that an intellectual basis exists for faith in Jesus Christ as the Son of God.[6] McDowell lists sixty-one Old Testament prophecies that he claims precisely foretold the coming of Jesus Christ as the Messiah.[7]

For example, consider Prophecy 1 (all these are exact quotations):[8]

PROPHECY

I will put enmity between you and the woman, and between your seed and her seed; he shall bruise your head, and you shall bruise his heel (Gen. 3:15, Revised Standard Version).

FULFILLMENT

But when the time had fully come, God sent forth his Son, born of a woman, born under the law (Gal. 4:4, Revised Standard Version).

I am not sure what the prediction is here; that Jesus was to be born of a woman?

McDowell often repeats himself. In Prophecies 14 and 32 he regards the statements in Luke 2:11, Mathew 22:43–45, Hebrews 1:3, Mark 16:19, and Acts 2:34–35 in which Jesus sits down on the right hand of God as a fulfillment of: "The Lord says to my lord: 'Sit at my right hand, till I make your enemies your footstool'" (Ps. 110:1, Revised Standard Version). McDowell certainly views biblical prophecy as something different than simple scientific prediction. I would not be too far off base to note that Jesus sitting on God's right hand has not been verified scientifically.

Each of the prophecies listed by McDowell is confirmed in no other place except in the Bible. We have no independent evidence that events actually took place as described—especially the ones happening in heaven. Before making the extraordinary claim that something supernatural occurred, simple common sense tells us that we must rule out the ordinary, far more plausible account that the events are fictional, written so as to conform to biblical prophecies.

For example, Prophecy 55 takes the opening words of one of David's Psalms, "My God, my God, why hast thou forsaken me" (Ps. 22:1a, King James Version) and sees this precisely fulfilled with Jesus' last words on the cross (Matt. 27:46). Which is the more plausible account: an extraordinary event in which a thousand years earlier David predicted the exact last words of the Messiah (although he does not identify them as such) or a perfectly ordinary one in which Matthew puts these words in Jesus' mouth when telling the story of the crucifixion? Or, perhaps Jesus really used these words, remembered from the Psalm.

Many of McDowell's examples have appeared frequently in Christian literature. Consider the prophecy of Jesus' coming: "But you, O Bethlehem Eph'rathah, who are little among the clans of Judah, from you shall come forth for me one who is to be ruler in Israel whose origin is from old, from ancient days" (Mic. 5:2, Revised Standard Version). We have no reason outside the New Testament to believe that Jesus was born in Bethlehem. History does not support Luke's Christmas story about a decree from Caesar Augustus that all the Roman world was required to go to their place of origin to be "taxed" (King James Version) or "enrolled" (Revised Standard Version). Surely such a vast undertaking would have been recorded. History does record a census affecting only Judea and not Galilee, but this took place in 6–7 CE, which conflicts with the fact that Jesus was supposedly born in the days of Herod, who died in 4 BCE.[9]

Similarly, we have no historical mention of a star lighting up the sky, although spectacular astronomical events such as comets and supernovae were frequently recorded in ancient times. And, surely there would be a record of Herod's slaughter of innocent children—had that really happened. The Jewish scholars Philo (c. 50) and Josephus (c. 93) described Herod as murderous and killing some family members to keep them from challenging his throne. Yet neither mentions the slaughter of the innocents.

Furthermore, Jesus was never the ruler of Israel. This aspect of the prophecy actually failed. And, he was never called "Immanuel" either, as the prophecy in Isaiah 7:14 foretold.

Perhaps one of the most important prophecies of the New Testament stands out like a sore thumb for its repeated appearance in the Gospels and gross failure to be fulfilled. In Matthew

16:28, 23:36, 24:34; Mark 9:1, 13:30; and Luke 9:27, Jesus tells his followers that he will return and establish his kingdom within a generation, before the listeners die. We are still waiting.

Lack of evidence from outside of scripture surrounds the most important tale of the New Testament—Jesus' crucifixion and resurrection. Christian literature is filled with claims that these events were foretold. But again we have nothing outside of the Gospels that rules out what is the more plausible account: the authors of the Gospels formulated the life and death of Jesus to conform to their conception of the Messiah of the Old Testament.

Many people say they believe because of the many eyewitnesses who said they saw Jesus walking after he was supposed to be dead. However, that testimony is only recorded in the Bible, second hand, and years after the fact. Eyewitness testimony recorded on the spot would still be open to question two thousand years after the fact. Eyewitness testimony recorded decades later is hardly extraordinary evidence.

Furthermore, eyewitness testimony recorded on the spot today is notoriously unreliable.[10] In a recent decade, sixty-nine convicts were released from prison, seven on death row, based on DNA evidence. In most cases, these people were convicted primarily on the basis of eyewitness testimony.

Now, as with the Christmas story, we might easily imagine that independent evidence could have been found. Matthew describes what happened at the death of Jesus: "And behold, the curtain of the temple was torn in two, from top to bottom; and the earth shook, and the rocks were split; the tombs we opened and many of the bodies of the saints who had fallen asleep were raised, and coming out of the tombs after his resurrection they went into the holy city and appeared to many"(Matt. 27:51–54, Revised Standard Version). Again, we have no record of these phenomenal events outside scripture. If they really happened as described, Philo, Josephus, or one of the many historians of the time would very likely have mentioned them.

The few mentions of "Christus" in the pagan literature, decades after Jesus' death, do not provide the needed confirmation. They read simply as factual reports on a new cult that was appearing in the empire. Considerable controversy still exists on the validity of various statements taken from the writings of Josephus, which seem to support specifics of the Gospel stories.[11] But, once again, these were written well after Jesus' death and were not firsthand observations. In short, despite the long list of Jewish and pagan scholars writing at that time,[12] there is no record of Jesus being tried by Pontius Pilate and executed—much less rising from the dead.

Christian apologist William Lane Craig cites the empty tomb as evidence for the risen Christ.[13] However, the Gospels are inconsistent in their description of this event, as the reader should check for herself. Simply compare the four accounts: Mark 16:1–8, Matthew 28:1–10, Luke 24:1–11, John 20:1–18. Even if we assume, for the sake of argument, that the story of the empty tomb is accurate, a much simpler explanation exists. Suppose you are on holiday in Paris and decide one morning to visit the tomb of Napoleon. You arise bright and early and find the tomb is empty. Would you conclude that the emperor had risen into heaven? Hardly. You would figure somebody took the body!

Since ancient times, many authors have commented on how the birth, life, death, and resurrection of Jesus as described in the Gospels are similar to those of savior-gods in various mystery cults and religions of the ancient world.[14] True, this remains an issue in much dispute. In his exhaustive study of the background of the early Church, Everett Ferguson warns us that many of these generalizations are fraught with methodological problems and that the similarities between the mystery religions and Christianity is exaggerated.[15] He admits, however, that much of that exaggeration came from Christian writers themselves. The Jesus story sure looks just like you would expect it to look if it were patterned after other god-men.[16]

Early Christian Church fathers such as Justin Martyr (d. 165), Tertullian (d. 225), and Irenaeus (d. 202) felt compelled to answer the pagan critics of the time who claimed the Jesus story was based on earlier traditions. The fathers claimed that the similarities were the work of the devil, who copied the Jesus story ahead of time to mislead the gullible.

Lacking any independent corroboration, we cannot take the New Testament as evidence for a single fulfilled Old Testament prophecy, much less sixty-one. The story of Jesus, as related in the Gospels, with all its unconfirmed miraculous happenings, is more plausibly explained as largely a fiction, written to not only conform to Judaic traditions but also to move Christianity beyond being a tribal religion. The story appealed to gentiles as well, with the incorporation of many of their god-man myths.[17]

This is not to say that the myth of Jesus is not based on a real person, although some scholars have tried to make that case.[18] That assumption is not necessary for the case against God. We have seen that the Gospels cannot be used as evidence for the success of various Old Testament predictions because we have no independent confirmation that those events ever occurred.

OLD TESTAMENT PROPHECIES

A similar conclusion can be drawn about Old Testament predictions of events within that document itself. On his Web site *Reasons to Believe*, physicist Hugh Ross lists a number of Old Testament prophecies that he claims were fulfilled. In addition to the predictions of a Messiah that have already been discussed, Ross lists several predictions where the predicted event occurs in the Old Testament. For example, quoting Ross, "One prophet of God (unnamed, but probably Shemiah) said that a future king of Judah, named Josiah, would take the bones of all the occultic priests (priests of the 'high places') of Israel's King Jeroboam and burn them on Jeroboam's altar" (1 Kings 13:2 and 2 Kings 23:15–18). This event occurred approximately three hundred years after it was foretold.[19] In his book *Bible Prophecy*, Tim Callahan notes, "1 Kings 13:2 refers to the northern kingdom as Samaria and since Israel was not referred to by the name of its capital until after it had fallen to the Assyrians in 721 BCE, the prophet crying out against Jeroboam's idolatry, which took place around 900 BCE, was inserted by Deuteronomists hundreds of years after the fact."[20] All Ross's examples, like those of McDowell, have no corroboration outside the Bible. Rather than making the extraordinary claim that something supernatural has happened and future events were foretold, the far more plausible, ordinary explanation is that the "prophecies" were inserted after the fact.

The Old Testament has numerous failures of prophecy as well. Here are just a few:

- Isaiah 17:1. Damascus is predicted to cease to be a city. In fact, Damascus is one of the oldest continuously inhabited cities.

- Jeremiah 49:33 predicts that Hazor will become an everlasting wasteland in which humans will never again dwell. The King James Bible says it will become inhabited by dragons. None of this has happened.

- Zechariah 10:11. The Nile is predicted to dry up. This has not yet happened.

- Ezekiel 29, 30. The land of Egypt will be laid waste by Nebuchadnezzar, all its people killed and rivers dried up. It will remain uninhabited for forty years. This did not happen.

Biblical scholars will argue endlessly about these issues, but it is not necessary to enter into their conflict. We need to seek evidence that will stand up under the kind of scrutiny scientists give to predictive claims of extraordinary events in any field. The fact is that no independent evidence exists that any biblical prophecy has been fulfilled, despite the insistent claims of apologists such as McDowell, Craig, and Ross.

PHYSICAL EVIDENCE

Events recorded two thousand years ago and longer by superstitious people accustomed more to mythological tales than objective observations cannot be taken literally. The scriptures look exactly as they would be expected to look if they were written without the deep insight of divine revelation.

But, again, it might have been different. Evidence might have been found that could not be dismissed as yet another myth in an ancient world filled with myths. Physical data, examined under the microscopes of modern science, could still provide for the type of verification of extraordinary claims that can be found in laboratories all over the world today.

In 1995 I walked into the mummy room in the Cairo Museum and gazed down on the earthly remains of Pharaoh Ramses II. I could see the hawkish facial features of the great king who reigned over Egypt for sixty-seven years, dying at age ninety-six in 1213 BCE—well over three thousand years ago! Thanks to the Rosetta Stone, discovered in 1799 during Napoleon's invasion of Egypt, and the numerous monuments (to himself) that Ramses had built during his reign, we know much about his life. While no doubt many of the exploits depicted on the walls of temples were exaggerated, we can be sure this man existed and that many of the details of his life are known.

A few months later, I visited a museum in Thessalonica, Greece. There I saw bones that I was told at the time belonged to King Philip II of Macedon (d. 336 BCE), the father of Alexander the Great. These remains had been discovered in Macedonia just a few years earlier, and my guide was a physicist involved in their dating. Since then, the remains have been reidentified as those of Alexander's half-brother, Philip III Arrhidaeus, who was assassinated in 317 BCE.[21]

So, here we have detailed, identifiable physical evidence for the existence of men who lived long before Jesus. Now, Jesus never ruled a worldly kingdom, but we have been led to believe he was a person of some local renown. In principle, we could have found either bones or some tablet from his time that confirmed Jesus' existence. The Shroud of Turin,[22] and the more recent discovery called the James Ossuary, have turned out to be likely forgeries.[23] They might not have been. And, in fact, these issues are still being debated. Perhaps someday a discovery will be made, especially since the region where Jesus lived is the most heavily excavated by archaeologists in the world.

For example, suppose some bones are found that can be identified as those of Jesus by means of accompanying physical evidence. This would disprove the doctrine that he rose bodily into heaven, which at least demonstrates that the hypothesis of Jesus' bodily resurrection is eminently falsifiable. Such a discovery would not ring the death knell of Christianity (although William Lane Craig seems to think so) where, today, most Christians think in terms of immaterial spirits as the entities that survive the grave and dwell in heaven or hell. A discovery of physical evidence for the Jesus of the Gospels would at least put to rest the doubts that have been expressed about the very existence of the carpenter from Galilee. If they showed signs of crucifixion, then that part of the biblical narrative would be confirmed. Following a suggestion of Richard Dawkins, we might even imagine that the DNA found in the bones did not represent those of an earthly human, providing confirmation of Jesus' otherworldly nature.

Of course this is all very hypothetical and not likely to ever happen. And the apologist can easily invent a host of reasons for why we have found no evidence. My point is simply that obtaining incontrovertible physical data confirming the validity of events as related in scriptures is not out of the realm of possibility. It could happen. Someday it may happen. So far, it has not.

UNEARTHING NOTHING

The lack of physical evidence does not prove necessarily that some person or event described in ancient chronicles is purely mythological. However, in the case of a number of biblical events, the absence of supporting physical evidence that *should* have been found with high probability allows us to draw a strong, scientific conviction that those events never took place.

Such is the case for several of the most important of Old Testament narratives that describe people and events surrounding the very foundations of Judaism, Christianity, and Islam. The details and references to the sources of data may be found in the startling book *The Bible Unearthed*, by Israel Finkelstein and Neil Asher Silberman.[24]

Perhaps the most important figure in the Old Testament (besides Yahweh himself) is Moses, who supposedly led the Jews out of captivity in Egypt and wandered the Sinai Desert for forty years. During his wanderings, according to the scripture, Moses often talked to God, obtained the Ten Commandments, and made a covenant between the people of Israel and Yahweh. With God's guidance, Moses finally brought his people to the Promised Land, which, as modern Israeli prime minister Golda Meir once remarked, was the only place in the Middle East with no oil.

Finkelstein and Silberman report that no recognizable archaeological evidence has been found of an Israelite presence in Egypt prior to the thirteenth century BCE, when most scholars believe the Exodus took place. This was around the time of Pharaoh Ramses II, whose remains I viewed in Cairo in 1995.[25]

According to the biblical account, six hundred thousand Jews participated in the escape from Egypt. Even if this is wildly exaggerated, Finkelstein and Silberman argue that some archaeological traces of their wandering should have been found by now. Despite extensive searching, "not a single campsite or sign of occupation from the time of Ramses II and his immediate predecessors and successors has even been identified in Sinai."[26]

Finkelstein and Silberman note that modern archaeological techniques are capable of tracing even the very meager remains of small bands of far more ancient hunter-gatherers and pastoral nomads all over the world. They say, "The conclusion—that the Exodus did not happen at the time and in the manner described in the Bible—seems irrefutable."[27]

Finkelstein and Silberman are part of a school of biblical archaeologists called "minimalists," who argue that many biblical accounts of early Israel have little, if any basis in factual data. They are opposed by the "maximalists," who still claim that Bible accounts are generally confirmed by archaeology. One highly respected scholar, William Dever, has tried to straddle the two positions. But he has to agree with the minimalists on the questions of Moses. "Absolutely no trace of Moses, or indeed of an Israelite presence in Egypt, has ever turned up. Of the Exodus and the wandering in the wilderness—events so crucial in the Biblical recitation of the 'mighty acts of God'—we have no evidence whatsoever. . . . Recent Israeli excavations at Kadesh-Barnea, the Sinai oasis where the Israelites are said to have encamped for forty years, have revealed an extensive settlement, but not so much as a potsherd earlier than the tenth century B.C."[28]

The minimalists also cast considerable doubt that the great battles in Canaan, which the Bible describes as happening after the death of Moses, actually occurred. The cities in the region were poor and unfortified at that time, and excavations show no signs of destruction. Jericho had no walls to come tumbling down at the blast of Joshua's trumpet. In fact, it was not even settled at that time, having been destroyed around 2400 BCE, some nine hundred years before Joshua's alleged conquest.[29]

In short, the stories of Moses and his immediate successors are surely myths. In science, the absence of evidence that is required by a hypothesis constitutes a falsification of the hypothesis. The hypothesis of a God who selected out a small desert tribe as his chosen people and communicated the law to them while they wandered the Sinai Desert is falsified by the absence of evidence required by that hypothesis.

After Moses and Abraham (who is also probably a mythological figure[30]), the next most important personages in the Old Testament are David and Solomon. They are described in the Bible as rulers of great wealth, presiding over the Golden Age of the briefly united kingdoms of Israel and Judea. Yet there is no mention of either king in Egyptian or Mesopotamian texts. No physical evidence has been found for David's conquests or his empire. Archaeological support for Solomon's great temple in Jerusalem or other building projects there and in other locales is nonexistent.[31]

At a recent meeting in Rome, archaeologist Niels Peter Lemche declared, "Archaeological data have now definitely confirmed that the empire of David and Solomon never existed."[32]

In 1993 a fragment of a black basalt monument was found at Tel Dan in northern Israel. It contained an inscription in Aramaic describing an assault on the northern kingdom of Israel by the king of Damascus around 835 BCE and his defeat of the "House of David." Some scholars now think this may be a forgery; in any case, it does not prove the existence of a united kingdom.[33]

Almost certainly, the Jewish kingdom was far more modest than described in the Bible, and the events surrounding David are probably as mythological as those of the lives of Abraham, Moses, and Jesus.

As you might expect, these conclusions continue to be hotly debated in the community of biblical scholars and archaeologists. Some maximalists have argued that the remains of

Solomon's temple and other signs of a Golden Age in Jerusalem have been wiped out by later building projects. However, the extensive excavations carried on in Jerusalem in modern times have yielded impressive finds from much earlier periods such as the Middle Bronze Age and Iron Age, which would have been covered by even more debris.[34]

In short, the hypothesis of a God who looked down with favor on a small desert tribe fifteen hundred to a thousand years before Christ, enabling them to forge a great, albeit short-lived empire, is falsified by the absence of data.

NOT EVEN REMOTELY HISTORIC

If the most important stories found in the Old and New Testaments are even remotely historic, then scientific evidence should exist for an escape of large numbers of Jews from Egypt in the thirteenth century BCE and forty years of wandering in the desert. It does not. Physical evidence should exist for great battles as the Israelites captured the land of Canaan, after returning to Canaan. It does not. Physical evidence should exist for a Golden Age in a combined kingdom of Israel and Judea around 1000 BCE and the Temple of Solomon. It does not.

Historical evidence should also exist for the extraordinary events reported to have occurred at the time of Jesus' birth. It does not. Historical evidence should exist for the extraordinary events reported to have occurred at the time of Jesus' death. It does not. From the absence of evidence that should exist in the scientific and historical record, we can conclude beyond a reasonable doubt that these extraordinary events did not take place as the Bible describes.

The Bible reads as an assembly of myths fashioned by ancient authors who had no concept of historical accuracy. Its description of the world reflects the scientific and historical knowledge of the age in which the manuscripts were composed.

The information and insights contained in scriptures and other revelations look just as they can be expected to look if there is no God who revealed truths to humanity that were recorded in sacred texts.

NOTES

1. Archer L. Gleason, *Encyclopedia of Bible Difficulties* (Grand Rapids, MI: Zondervan, 2001), p. 23.

2. See, for example, Gen. 1:6–8; Chron. 16:30; Ps. 93:1, 96:10, 104:5; Isa. 45:18. Isa. 40:22 says Earth is a "circle." Note that a circle is flat. Both the King James and Revised Standard versions have been consulted here.

3. For an attempt to make the Bible creation story consistent with science, see Gerald L. Schroeder, *Genesis and the Big Bang: The Discovery of the Harmony between Modern Science and the Bible* (New York: Bantam Books, 1992); *The Science of God: The Convergence of Scientific and Biblical Wisdom* (New York: Broadway Books, 1998); *The Hidden Face of God: How Science Reveals the Ultimate Truth* (New York: Free Press, 2001). For reviews, see Victor J. Stenger, "Fitting the Bible to the Data," *Skeptical Inquirer* 23 no. 4 (1999): 67, also online at *Secular Web*, http://www.infidels.org/library/modern/vic_stenger/schrev.html (accessed December 13, 2004), pp. 165–70; and Mark Perakh, "Not a Very Big Bang about Genesis" (December 2001), online at *Talk Reason*, http://www.talkreason.org/articles/schroeder.cfm (accessed December 15, 2004).

4. Thanks to Brent Meeker for rewriting this for me in Biblespeak.

5. Josh McDowell, *Evidence That Demands a Verdict* (San Bernardino, CA: Here's Life Publishers, 1972, 1979). Quotations from paperback version of revised edition.

6. For a chapter-by-chapter critique, see Jeffery Jay Lowder, ed., "The Jury Is In: The Ruling on McDowell's 'Evidence,'" online at *Secular Web*, http://www.infidels.org/library/modern/jeff_lowder/jury/ (accessed January 14, 2005). The essays date from 1997 to 1999.

7. McDowell, *Evidence That Demands a Verdict*, pp. 141–66.

8. Ibid., p. 145.

9. Randel Helms, *Gospel Fictions* (Amherst, NY: Prometheus Books, 1988), p. 59.

10. Elizabeth F. Loftus, *Eyewitness Testimony* (Cambridge, MA: Harvard University Press, 1996).

11. Everitt Ferguson, *Background of Early Christianity*, 3d ed. (Grand Rapids, MI: W. B. Eerdmans, 2003), p. 488.

12. Timothy Freke and Peter Gandy, *The Jesus Mysteries: Was the "Original Jesus" a Pagan God?* (New York: Harmony Books, 1999), p. 133.

13. William Lane Craig, "The Historicity of the Empty Tomb of Jesus," *New Testament Studies* 31 (1985): 39–67, http://www.leaderu.com/offices/billcraig/docs/tomb2.html (accessed January 4, 2005).

14. Freke and Gandy, *The Jesus Mysteries*.

15. Ferguson, *Background of Early Christianity*, pp. 297–300.

16. See, for example, Philostratus, *The Life of Apollonius of Tyana*, quoted in Helms, *Gospel Fictions*, p. 9.

17. Helms, *Gospel Fictions*.

18. Joseph R. Hoffmann and Gerald A. Larue, eds., *Jesus in History and Myth* (Amherst, NY: Prometheus Books, 1986); G. A. Wells, *The Historical Evidence for Jesus* (Amherst, NY: Prometheus Books, 1988); Earl Doherty, *The Jesus Puzzle: Did Christianity Begin with a Mythical Christ?* (Ottawa: Canadian Humanist Publications, 1999).

19. Hugh Ross, "Fulfilled Prophecy," http://www.reasons.org/resources/apologetics/prophecy.shtml (accessed January 14, 2005). Original dating 1975, revised August 22, 2003.

20. Tim Callahan, *Bible Prophecy: Failure or Fulfillment* (Altadena, CA: Millennium Press, 1997), p. 47.

21. Angela M. H. Schuster, "Not Phillip II of Macedon," *Archaeology* (April 20, 2000), http://www.archaeology.org/online/features/macedon/ (accessed December 26, 2004).

22. Joe Nickell, *Inquest on the Shroud of Turin* (Amherst, NY: Prometheus Books, 1987).

23. Ibid., "Bone (Box) of Contention: The James Ossuary," *Skeptical Inquirer* 27, no. 2 (March/April 2003): 19–22. A complete set of scholarly essays on the James Ossuary is online at *Bible and Interpretation*, http://www.bibleinterp.com/articles/James_Ossuary_essays.htm (accessed December 25, 2004).

gian Ronald Sider lamented: "Scandalous behavior is rapidly destroying American Christianity. By their daily activity, most 'Christians' regularly commit treason. With their mouths they claim that Jesus is Lord, but with their actions they demonstrate allegiance to money, sex, and self-fulfillment."[5] Sider continues,

> The findings in numerous national polls conducted by highly respected pollsters like The Gallup Organization and The Barna Group are simply shocking. "Gallup and Barna," laments evangelical theologian Michael Horton, "hand us survey after survey demonstrating that evangelical Christians are as likely to embrace lifestyles every bit as hedonistic, materialistic, self-centered, and sexually immoral as the world in general." Divorce is *more* common among "born-again" Christians than in the general American population. Only 6 percent of evangelicals tithe. White evangelicals are the *most* likely people to object to neighbors of another race. Josh McDowell has pointed out that the sexual promiscuity of evangelical youth is only a little less outrageous than that of their nonevangelical peers.

COMMON STANDARDS

It is not my purpose in this chapter to say how humans ought to behave. Rather I am acting as a scientist, observing how they do behave and asking what those observations tell us about the truth or falsity of the God hypothesis. In this regard, I reject the notion that science has nothing to say about morality.

Preachers tell us that any universal moral standards can only come from one source—their particular God. Otherwise standards would be relative, depending on culture and differing across cultures and individuals. The data, however, indicate that the majority of human beings from all cultures and all religions or no religion agree on a common set of moral standards. While specific differences can be found, universal norms do seem to exist. Anthropologist Solomon Asch has observed, "We do not know of societies in which bravery is despised and cowardice held up to honor, in which generosity is considered a vice and ingratitude a virtue."[6]

While we live in a society of law, much of what we do is not constrained by law but performed voluntarily. For example, we have many opportunities to cheat and steal in situations where the chance of being caught is negligible, yet most of us do not cheat and steal. While the Golden Rule is not usually obeyed to the letter, we generally do not try to harm others. Indeed, we are sympathetic when we see a person or animal in distress and take action to provide relief. We stop at auto accidents and render aid. We call the police when we witness a crime. We take care of children, aged parents, and others less fortunate than us. We willingly take on risky jobs, such as in the military or public safety, for the protection of the community.

That stealing from members of your own community is immoral requires no divine revelation. It is revealed by a moment's reflection on the type of society that would exist if everyone stole from one another. If lying were considered a virtue instead of truth-telling, communication would become impossible. Mothers have loved their children since before mammals walked the earth—for obvious evolutionary reasons. The only precepts unique to religion are those telling us to not to question their dogma.

Of course, not everyone agrees on every moral issue. These disagreements can be very pronounced, especially within specific religious communities where the same scriptural readings are often used to justify contradictory actions.

For example, consider the opposing interpretations of the commandment against killing found within the Christian community. Conservative Protestants interpret this commandment as prohibiting abortion, stem cell research, and removing life support systems from the incurable, among other actions. However, they do not view capital punishment as prohibited, pointing to the biblical prescription of an eye for an eye. Catholics and liberal Christians, on the other hand, generally interpret the commandment as forbidding capital punishment. But Catholics oppose while liberals allow abortion, the removal of life support, and stem cell research. In all these cases, the Bible is evidently ambiguous.

As philosopher Theodore Schick Jr. points out, both sides of the abortion debate believe murder is immoral. Where they disagree is on the nature of a fetus—whether or not it is the sort of entity that can be murdered. In other words, moral disagreements are often not about what is good or bad but about some other aspect of reality.[7]

So how do Christians decide what is right or wrong? While they may look at the Bible, how they interpret what they read must depend on ideals that they have already developed from some other source.

NOBLE IDEALS

The Judeo-Christian and Islamic scriptures contain many passages that teach noble ideals that the human race has done well to adopt as norms of behavior and, where appropriate, to codify into law. But without exception, the fact that these principles developed in earlier cultures and history indicates that they were adopted by—rather than learned from—religion. While it is fine that religions preach moral precepts, they have no basis to claim that these precepts were authored by their particular deity or, indeed, any deity at all.

Perhaps the primary principle upon which to live a moral life is the Golden Rule: *"Do unto others, as you would have them do unto you."* In our Christian-dominated society in the West, most people assume that this was an original teaching of Jesus from the Sermon on the Mount. For some reason, their preachers, who surely know better, perpetuate this falsehood. In fact, Jesus himself made no such claim. Here's what he actually said, according to the Gospel: "So, whatever you wish that men would do to you, do so to them; for this is the law of the prophets" (Matt. 7:12, Revised Standard Version). Indeed, the phrase "Love thy neighbor as thyself" appears in Leviticus 19:18, written a thousand years before Christ.

But the Golden Rule is not the exclusive property of a small desert tribe with a high opinion of itself. Here are some other, independent sources showing that the Golden Rule was already a widespread teaching well before Jesus:

- In *The Doctrine of the Mean 13*, written about 500 BCE, Confucius says, "What you do not want others to do to you, do not do to others."
- Isocrates (c. 375 BCE) said, "Do not do to others what would anger you if done to you by others."
- The Hindu *Mahabharata*, written around 150 BCE, teaches, "This is the sum of all true righteousness: deal with others as thou wouldst thyself be dealt by."[8]

In the Sermon on the Mount, Jesus also urged his listeners, "Do not resist one who is evil. But if any one strikes you on the right cheek, turn to him the other also" (Matt. 5:39, Revised Standard Version) and "You have heard that it was said, 'You shall love your neighbor and hate your enemy.' But I say to you, Love your enemies and pray for those who persecute you" (Matt. 5:43–44, Revised Standard Version).

Again, these are generally regarded as uniquely Christian sentiments. But the call to "love your enemies" precedes Jesus and does not even appear in the Old Testament:[9]

- I treat those who are good with goodness. And I also treat those who are not good with goodness. Thus goodness is attained. I am honest with those who are honest. And I am also honest with those who are dishonest. Thus honesty is attained (Taoism. *Tao Te Ching* 49).

- Conquer anger by love. Conquer evil by good. Conquer the stingy by giving. Conquer the liar by truth (Buddhism. *Dhammapada* 223).
- A superior being does not render evil for evil; this is a maxim one should observe; the ornament of virtuous persons is their conduct. One should never harm the wicked or the good or even criminals meriting death. A noble soul will ever exercise compassion even towards those who enjoy injuring others or those of cruel deeds when they are actually committing them—for who is without fault? (Hinduism. *Ramayana*, Yuddha Kanda 115).

No original moral concept of any significance can be found in the New Testament. In the early twentieth century, historian Joseph McCabe noted: "The sentiments attributed to Christ are . . . already found in the Old Testament. . . . They were familiar in the Jewish schools, and to all the Pharisees, long before the time of Christ, as they were familiar in all the civilizations of the earth—Egyptian, Babylonian, and Persian, Greek and Hindu."[10]

As with the Bible, the Qur'an contains many sentiments that most of us would classify as commendable. It tells Muslims to be kind to their parents, not to steal from orphans, not to lend money at excess interest, to help the needy, and not to kill their children unless it is necessary.

But, again, these are not original moral principles. In the scriptures and other teachings of the great monotheisms we find a repetition of common ideals that arose during the gradual evolution of human societies, as they became more civilized, developed rational thinking processes, and discovered how to live together in greater harmony. The evidence points to a source other than the revelations claimed in these scriptures.

THE GOOD SOCIETY

Not only personal behavior but also societal behavior is supposedly regulated by God. But, once again, we can find no evidence for
this. One of the prevailing myths in modern America is that the
nation was founded on "Christian principles." However, the
United States Constitution is a secular document that contains no
reference to God, Jesus, Christianity, salvation, or any other religious teaching. Most of the early presidents were not fervent Christians and based their commitments to freedom, democracy, and
justice on Enlightenment philosophy rather than biblical sources.

We often hear, especially from American politicians, that our
legal system is founded on the Ten Commandments. Attempts
have been made to display the Ten Commandments in public facilities such as courthouses, which the courts have so far disallowed.
But, we need to read what the commandments actually say.

Since there are several versions, let me present a simplified
wording with religious language omitted:[11]

The Ten Commandments

1. Have no other gods before me.
2. Make no images of anything in heaven, Earth, or the sea,
   and do not worship or labor for them.
3. Do not use the name of your God in vain.
4. Do no work on the Sabbath.
5. Honor your parents.
6. Do not kill.
7. Do not commit adultery.
8. Do not steal.
9. Do not give false testimony against another.
10. Do not desire another's wife or anything that belongs to
    another.

Only commandments 6, 8, and 9 (the numbering is different for Catholics and Protestants) can be found in the laws of any modern nation. Killing, stealing, and perjury are illegal—except when done by the government. While adultery is normally considered immoral, it is not generally illegal.

The Old Testament contains many examples of killings performed under God's orders. The only way this can be reconciled with commandment 6 is to assume that the proscription against killing must be restricted, say, only to your particular tribe rather than all humanity.

And, how many believers realize they are breaking commandment 2 every time they take a photograph or draw a picture? How many would stop if that were pointed out to them?

The restrictions imposed by the Ten Commandments can be found in other civilizations predating the time of Moses. Furthermore, it is clear from the above list that most of these restrictions are irrelevant to modern life and hardly form the basis for any existing legal system. Indeed, the Code of Hammurabi (c. 1780 BCE) represents a considerably more significant historical step in the development of laws of justice, containing not merely 10 but 282 detailed commandments.[12] Perhaps these should be displayed on courthouse steps.

Or, another option would be the Laws of Solon. Solon (d. 558 BCE) was an Athenian who is regarded as the founder of Western democracy and the first man in Western history to record a written constitution. That constitution eliminated birth as a basis for government office and created democratic assemblies open to all male citizens, such that no law could be passed without the majority vote of all. (Equal rights for women were still a long way off.) American democracy owes far more to Solon than the crude rules of the Hebrews.[13]

Christendom and Islam have a long history of authoritarianism with little disposition toward individual freedom and justice. Nowhere in the Bible can you discover the principles upon which modern democracies and justice systems are founded.

Slavery provides another example where the Bible hardly forms a model for our modern free societies. The Old Testament not only condones slavery but actually regulates its practice:

When you buy a Hebrew slave, he shall serve six years, and in the seventh he shall go out free, for nothing. (Exod. 21:2, Revised Standard Version)

If his master gives him a wife and she bears him sons or daughters, the wife and her children shall be her master's and he shall go out alone. (Exod. 21:4, Revised Standard Version)

Jesus had many opportunities to disavow slavery. He never did. St. Paul reaffirms the practice: "Bid slaves to be submissive to their masters and to give satisfaction in every respect" (Titus 2:9).

Prior to the Civil War, the Bible was widely used to justify slavery in the United States. Baptist leader and slave owner Richard Furman (d. 1825) laid the foundation for the biblical arguments that would be made in support of slavery leading up to the Civil War. While president of the State Baptist Convention, Furman wrote to the governor of South Carolina, "The right of holding slaves is clearly established in the Holy Scriptures, both by precept and example."[14] Furman University in Greenville, South Carolina, founded in 1826, was named for Richard Furman; his writings can be found in its archives.

Another prominent churchman, Alexander Campbell (d. 1866) wrote, "There is not one verse in the Bible inhibiting slavery, but many regulating it. It is not then, we conclude, immoral."[15] It is to be noted that Campbell declared himself against slavery, so once again we have a Christian following his own conscience despite what the scriptures say.

Jefferson Davis, president of the Confederacy, claimed to

follow what the scriptures said: "[Slavery] was established by decree of Almighty God . . . it is sanctioned in the Bible, in both Testaments, from Genesis to Revelation."[16]

While Christians in the South held onto their slaves as long as they could, secular humanist Richard Randolph of Virginia began freeing his in 1791.[17] Popes and other fathers of the Catholic Church owned slaves as late as 1800. Jesuits in colonial Maryland and nuns in Europe and Latin America owned slaves. The Church did not condemn slavery until 1888, after every Christian nation had abolished the practice.[18]

Distinguished Catholic scholar John T. Noonan Jr. points out that the Church has traditionally denied that it has made any changes in the moral teachings of Jesus and the apostles.[19] Slavery and other examples he presents amply illustrate that the Church's teaching does indeed change with the times.

Now, the campaign to end slavery in the United States and elsewhere was led by Christians, to their everlasting credit. However, the abolitionists clearly were not guided by the literal words of scripture but by their own interpretations and innate senses of a higher good.

Finally, let me just briefly mention the historical oppression of women. St. Paul said, "Wives, be subject to your husbands, as to the Lord. For the husband is the head of the wife as Christ is the head of the church, his body, and is himself its Savior" (Eph. 5:22–23, Revised Standard Version). Western societies finally have begun to recognize the irrationality and injustice of treating women as lesser human beings, providing a clear, recent example of how our notions of right and wrong evolve independent of and often contrary to religious teachings.

HOLY HORRORS

The Old Testament is filled with atrocities committed in the name of God. These are rarely mentioned in Sunday school, but anyone can pick up a Bible and read them for herself. I will just mention some of the worst: "Now therefore, kill every male among the little ones, and kill every woman that hath known man by lying with him. But all the young girls who have not known a man by lying with him, keep alive for yourselves" (Num. 31:17–18, Revised Standard Version).

At another time, Moses orders three thousand men put to the sword on God's authority: "And he said to them, 'Thus says the Lord God of Israel, "Put every man his sword on his side, and go to and fro from gate to gate throughout the camp, and slay every man his companion, and every man his neighbor"'" (Exod. 32:27, Revised Standard Version).

Most Christians dismiss this and other biblical carnage as anachronistic and imagine such orders were eliminated with the coming of Jesus. However, in the New Testament, Jesus frequently reaffirms the laws of the prophets: "Think not that I have come to abolish the law, or the prophets: I have come not to abolish them but to fulfil them" (Matt. 5:17, Revised Standard Version). The theist may respond that the above quotation is not a law but merely the report of an event, but the stories of the Bible are supposed to provide guides to proper behavior.

Christians like to pride themselves on their "family values" and their desire for peace in the world. No doubt, most are devoted to their families and are upright members of society. But they fail to remember that Jesus said: "Do not think that I have come to send peace on earth: I have not come to send peace, but a sword. For I have come to set a man against his father, a daughter against her mother, and a daughter-in-law against her mother-in-law; and a man's foes will be those of his own household. He who loves father or mother more than me is not worthy of me; and he who loves son or daughter more than me is not worthy of me" (Matt. 10:34–37, Revised Standard Version).

The history of Christendom abounds with violence sanctioned by the Church and thereby defined as divinely inspired "good." This divine inspiration is not limited to scripture but continually available to the specially anointed. Pope Urban II (d. 1099) assured the medieval knights of the Crusades that the killing of infidels was not a sin. And this did not apply just to Muslims in the Holy Land. The Cathar faith in southern France, which was apparently based on the notion of dual gods that appeared earlier in Zoroastrianism and Manichaeism,[20] was brutally suppressed in the Albigensian Crusade in the thirteenth century. When the besieged Cathar city of Beziers fell in 1209, soldiers reportedly asked their papal adviser how to distinguish the faithful from the infidel among the captives. He recommended: "Kill them all. God will know his own." Nearly twenty thousand were slaughtered—many first blinded, mutilated, dragged behind horses, or used for target practice.[21]

Incidentally, until recently the term *crusade* was used to refer to a Christian holy war, the equivalent of the Islamic *jihad*. Lloyd George's book of speeches given during his stint as British prime minister during the First World War was called *The Great Crusade*. General Dwight Eisenhower's memoir of the Second World War was called *Crusade in Europe*. The term *crusade* only fell into disuse recently, when shortly after September 11, 2001, President George W. Bush used it to refer to the war on terrorism and was warned off by his advisers because of its negative connotation for Muslims.[22] Of course, the Muslim terrorists themselves felt they were obeying God's command to engage in jihad.

The Qur'an is as bloodthirsty as the Old Testament. Numerous references can be found for the horrible fate that awaits nonbelievers. However, it is Allah himself who generally metes out that punishment: "Lo! Those who disbelieve Our revelations, We shall expose them to the Fire. As often as their skins are consumed We shall exchange them for fresh skins that they may taste the torment. Lo! Allah is ever Mighty, Wise" (Qur'an 4:56). Muslims are enjoined to kill infidels wherever they find them, but only those who initiate hostilities:

Fight in the way of Allah against those who fight against you, but begin not hostilities. Lo! Allah loveth not aggressors. Do not fight wars of aggression. And slay them wherever ye find them, and drive them out of the places whence they drove you out, for persecution is worse than slaughter. And fight not with them at the Inviolable Place of Worship until they first attack you there, but if they attack you (there) then slay them. Such is the reward of disbelievers. But if they desist, then lo! Allah is Forgiving, Merciful. And fight them until persecution is no more, and religion is for Allah. But if they desist, then let there be no hostility except against wrong-doers. (Qur'an 2:190–193)

Of course, in every religion there are a few fanatics who follow to the letter what they regard as God's will:

- Yigal Amir, who assassinated Israeli prime minister Yitzhak Rabin in 1995, was an extremely religious Jew who stated in court, "Everything I did, I did for God."[23]
- Paul Hill, who murdered abortion provider Dr. John Britton in Florida in 1994, made the following statement just before his execution in 2003: "I feel very honored that they are most likely going to kill me for what I did. I'm certainly, to be quite honest, I'm expecting a great reward in heaven for my obedience."[24]
- Mohammed Bouyeri, the Muslim extremist who killed Dutch filmmaker Theo van Gogh in 2004, declared in his trial, "What moved me to do what I did was purely my faith. . . . I was motivated by the law that commands me to cut off the head of anyone who insults Allah and his prophet."[25]

But, thankfully, they are the exception. Furthermore, each of these fanatics would be hard-pressed to demonstrate where exactly in their scriptures were they commanded to commit their dreadful acts.

Of course, no one of conscience today would think it moral to kill everyone captured in battle, saving only the virgin girls for their pleasure. Few modern Christians take the commands of the Bible literally. While they claim to appeal to scriptures and the teachings of the great founders and leaders of their faiths, they pick and choose what to follow—guided by some personal inner light. And this is the same inner light that guides nonbelievers.

AN INNER LIGHT

If God does not define what is good, who does? How are theists supposed to decide what is good?

Most do not go so far as to say that they hear it directly from God. While they claim to appeal to scriptures and the teachings of the great founders and leaders of their faiths, they pick and choose what to follow—guided by some personal inner light.

A good example is the Catholic community in the United States. Shortly after the death of Pope John Paul II in 2005, the *New York Times* reported:

> The roughly 65 million Catholics in the United States no longer have as distinctive an identity as they did a generation ago, and as they assimilated more thoroughly into American society, their views on social and moral issues came to mirror those of other Americans.

> "Catholics as a whole occupy the mainstream of American life, when 50 or 60 years ago, they were on the periphery of society," said John Green, director of the Ray C. Bliss Institute of Applied Politics at the University of Akron in Ohio and an expert on religion and politics.

As a result, the Vatican's teachings on a number of subjects, including contraception, the ordination of women, and homosexuality, are out of step with the beliefs and lifestyles of most American Catholics. But the Americans mostly find a way to stay in their faith by adhering to values most important to them and quietly ignoring those they disagree with.[26]

The Bible is not clear on what may be killed and what may not be. It does not explicitly sanction or forbid the killing of a fetus or stem cell. And, it certainly sanctions the killing of enemies, specifically those who do not worship Yahweh.

In all these cases, believers clearly read the Bible to find support for moral principles that they have already developed from some other source.

Christians draw Jesus Christ in their own image. As philosopher George Smith explains, "Because of the theological obligation to endorse the precepts of Jesus, Christian theologians have a strong tendency to read their own moral conviction into the ethics of Jesus. Jesus is made to say what theologians think he *should* have said."[27]

Philosopher Walter Kaufmann agrees, "Most Christians gerrymander the Gospels and carve an idealized self-portrait out of the texts: Pierre van Passen's Jesus is a socialist, Fosdick's is a liberal, while the ethic of Reinhold Niebuhr's Jesus agrees, not surprisingly, with Niebuhr's own."[28] As George Bernard Shaw commented, "No man ever believes that the Bible means what it says. He is always convinced that it says what he means."[29]

Every time a theologian reinterprets Moses, Jesus, or Muhammad, he further reinforces my crucial point: we humans decide what is good by standards lying outside the scriptures.

Believers are guided by their consciences in deciding for themselves what is right and wrong, just as are nonbelievers. The basic notions of good and evil that we all share—believers and nonbelievers—are, for the most part, common and universal. Psychological tests indicate that there are no significant differences in the moral sense between atheists and theists.[30]

In short, the empirical facts indicate that most humans are moral animals whose sense of right and wrong conflicts with many of the teachings of the great monotheistic religions. We can safely conclude they did not originate at that source.

NATURAL MORALITY

If human morals and values do not arise out of divine command, then where do they come from? They come from our common humanity. They can be properly called humanistic.[31]

A considerable literature exists on the natural (biological, cultural, evolutionary) origins of morality.[32] Darwin saw the evolutionary advantage of cooperation and altruism. Modern thinkers have elaborated on this observation, showing in detail how our moral sense can have arisen naturally during the development of modern humanity.

We can even see signs of moral, or protomoral behavior in animals. Vampire bats share food. Apes and monkeys comfort members of their group who are upset and work together to get food. Dolphins push sick members of a pod to the surface to get air. Whales will put themselves in harm's way to help a wounded member of their group. Elephants try their best to save injured members of their families.[33]

In these examples we glimpse the beginnings of the morality that advanced to higher levels with human evolution. You may call animal morality instinctive, built into the genes of animals by biological evolution. But when we include cultural evolution as well, we have a plausible mechanism for the development of human morality—by Darwinian selection.

It seems likely that this is where we humans have learned our sense of right and wrong. We have taught it to ourselves.

THE MORAL ARGUMENT

Since Thomas Aquinas, theologians have claimed that the very fact that humans have a moral conscience can be taken as evidence for the existence of God:

> There must be something which is to all beings the cause of their being, goodness and every other perfection: and this we call God.
>
> —Thomas Aquinas[34]

Contemporary Christian apologist William Lane Craig puts it this way, "If we can in some measure be good, then it follows that God exists."[35]

However, I have turned that argument on its head. The very fact that humans have a common moral conscience can be taken as evidence against the existence of God.

As we have seen from an examination of the empirical evidence, God cannot be the source of commonly accepted human morals and values. If he were, then we would expect to see evidence in the superior moral behavior of believers compared to nonbelievers. Even if you deny that any discrepancy exists between the behavior of believers and what is taught in their scriptures, the empirical fact that nonbelievers show themselves to be no less virtuous provides strong evidence that morals and values come from humanity itself. Observable human and societal behaviors look just as they can be expected to look if there is no God.

NOTES

    1. Phillip E. Johnson, *Darwin on Trial* (Downers Grove, IL: Inter-Varsity Press, 1991).

    2. According to a March 5, 1997, letter to Rod Swift from Denise

Golumbaski, research analyst, Federal Bureau of Prisons, online at http://www.holysmoke.org/icr-pri.htm (accessed February 5, 2006).

3. Ruth Miller, Larry S. Miller, and Mary R. Langenbrunner, "Religiosity and Child Sexual Abuse: A Risk Factor Assessment," *Journal of Child Sexual Abuse* 6, no. 4 (1997): 14–34.

4. Michael Franklin and Marian Hetherly, "How Fundamentalism Affects Society," *Humanist* 57 (September/October 1997): 25.

5. Ronald J. Sider, "The Scandal of the Evangelical Conscience," *Christianity Today* 11, no. 1 (January/February 2005): 8, http://www.christianitytoday.com/bc/2005/001/3.8.html (accessed March 22, 2005).

6. Solomon Asch, *Social Psychology* (Englewood Cliffs, NJ: Prentice-Hall, 1952), pp. 378–79.

7. Theodore Schick Jr., "Is Morality a Matter of Taste? Why Professional Ethicists Think That Morality Is *Not* Purely Subjective," *Free Inquiry* 18, no. 4 (1998): 32–34.

8. For other historical statements of the Golden Rule, see Michael Shermer, *The Science of Good & Evil: Why People Cheat, Gossip, Care, Share, and Follow the Golden Rule* (New York: Times Books, 2004), p. 23.

9. Thanks to Eleanor Binnings for providing these quotations.

10. Joseph McCabe, *The Sources of Morality of the Gospels* (London: Watts and Co., 1914), p. 209, as quoted in George Smith, *Atheism: The Case Against God* (Amherst, NY: Prometheus Books, 1989), p. 317.

11. Richard Carrier, "The Real Ten Commandments," Internet Infidels Library (2000), http://www.infidels.org/library/modern/features/2000/carrier2.html (accessed August 14, 2005).

12. The text of the Code of Hammurabi, translated by L. W. King, online at http://www.fordham.edu/halsall/ancient/hamcode.html (accessed April 3, 2005). Commentaries by Charles F. Horne (1915) and the *Encyclopaedia Brittanica* entry, 11th ed. (1910), written by Claude Hermann Walter Johns, also can be found at this site.

13. Carrier, "The Real Ten Commandments."

14. Richard Furman, "Exposition of the View of the Baptists Relative to the Colored Population of the United States to the Governor of South Carolina 1822," transcribed by T. Lloyd Benson from the original text in the South Carolina Baptist Historical Collection, Furman Uni-

versity, Greenville, South Carolina. Available at http://alpha.furman
.edu/~benson/docs/rcd-fmn1.htm (accessed December 1, 2004), p. 6.

15. Alexander Campbell, "Our Position to American Slavery—No.
V," *Millennial Harbinger*, ser. 3, vol. 2 (1845): 193.

16. Jefferson Davis, "Inaugural Address as Provisional President of
the Confederacy," Montgomery, AL, February 18, 1861, *Confederate States
of America Congressional Journal* 1 (1861): 64–66; quoted in Dunbar Rowland, *Jefferson Davis's Place in History as Revealed in His Letters, Papers, and
Speeches*, vol. 1 (Jackson, MS: Torgerson Press, 1923), p. 286.

17. Melvin Patrick Ely, *Israel on the Appomattox: A Southern Experiment in Black Freedom from the 1790s through the Civil War* (New York:
Alfred A. Knopf, 2005).

18. John T. Noonan Jr., *A Church That Can and Cannot Change: The
Development of Catholic Moral Teaching* (Notre Dame, IN: University of
Notre Dame Press, 2005).

19. Ibid.

20. Jean Markale, *Montségur and the Mystery of the Cathars*, trans.
Jon Graham (Rochester, VT: Inner Traditions, 2003).

21. For this and other tales of atrocities in the name of religion, see
James A. Haught, *Holy Horrors: An Illustrated History of Religious Murder
and Madness* (Amherst, NY: Prometheus Books, 1990).

22. Joan Acocella, "Holy Smoke; What Were the Crusades Really
About?" *New Yorker*, December 13, 2004.

23. CNN Report, March 27, 1996, http://www.cnn.com/WORLD/
9603/amir_verdict/ (accessed December 9, 2004).

24. Associated Press, September 2, 2003, http://www.fadp.org/
news/TampaBayOnline-20030903.htm (accessed December 9, 2004).

25. Trial statement, Associated Press, July 12, 2005, http://www
.guardian.co.uk/worldlatest/story/0,1280,-5136448,00.html (accessed
July 20, 2005).

26. Dean E. Murphy and Neela Banjeree, "Catholics in U.S. Keep
Faith, but Live with Contradictions," *New York Times*, April 11, 2005.

27. Smith, *Atheism: The Case Against God*, p. 313.

28. Walter Kaufmann, *The Faith of a Heretic*, paperback ed. (New
York: Doubleday, 1963), p. 216.

29. From a *Saturday Review* article, April 16, 1895.

30. Marc Hauser and Peter Singer, "Morality without Religion," *Free Inquiry* 26, no. 1 (December 2005/January 2006): 18–19.

31. Paul Kurtz, *Forbidden Fruit: The Ethics of Humanism* (Amherst, NY: Prometheus Books, 1988).

32. Robert Axelrod, *The Evolution of Cooperation* (New York: Basic Books, 1984); Richard D. Alexander, *The Biology of Moral Systems* (Hawthorne, NY: Aldine de Gruyter, 1987); Robert Wright, *The Moral Animal: Why We Are the Way We Are: The New Science of Evolutionary Psychology* (New York: Vintage Books, 1994); Frans B. M. de Wall, *Good Natured: The Origins of Right and Wrong in Humans and Other Animals* (Cambridge, MS: Harvard University Press, 1996); Larry Arnhart, *Darwinian Natural Right; The Biological Ethics of Human Nature* (Albany, NY: State University of New York Press, 1998); Leonard D. Katz, ed., *Evolutionary Origins of Morality: Cross-Disciplinary Perspectives* (Bowling Green, OH: Imprint Academic, 2000); Jessica C. Flack and Frans B. M. de Wall, "'Any Animal Whatever' Darwinian Building Blocks of Morality in Monkeys and Apes," *Journal of Consciousness Studies* 7, nos. 1–2 (2000): 1–29; Donald M. Broom, *The Evolution of Morality and Religion* (Cambridge: Cambridge University Press, 2003); Shermer, *The Science of Good & Evil*.

33. Shermer, *The Science of Good & Evil*, pp. 26–31.

34. Thomas Aquinas, Fourth Way in *Summa Theologica*.

35. William Lane Craig, "The Absurdity of Life without God," http://www.hisdefense.org/audio/wc_audio.html (accessed March 9, 2004).

Chapter 8

THE ARGUMENT FROM EVIL

With or without religion, good people can behave well and bad people can do evil; but for good people to do evil—that takes religion.

—Steven Weinberg[1]

THE PROBLEM OF EVIL

Although the ancient problem of evil is usually discussed in philosophical and theological rather than scientific terms, it is so important to the debate over the existence of God that I have included a discussion in this chapter for the sake of completeness. Besides, we might argue that a scientific element does enter in the empirical fact that very bad things, such as gratuitous suffering, happen in the world.

The problem of evil can be formally stated as follows:

1. If God exists, then the attributes of God are consistent with the existence of evil.
2. The attributes of God are not consistent with the existence of evil.
3. Therefore, God does not and cannot exist.[2]

The primary attributes that apply here I have designated as "3O"—omnibenevolence, omnipotence, and omniscience. Recall that these attributes were not included explicitly in what I called the "scientific God model" (see chap. 1) since the arguments presented in this book are not limited to a god with these qualities. Nevertheless, the traditional God of the great monotheisms is assumed to have the 3O attributes, which leads to an enormous logical difficulty that theologians have wrestled with over centuries without success. How can the 3O God be reconciled with the existence of evil?

The attempt to defend the notion of a God of infinite goodness, power, and wisdom in light of the undeniable existence of pain and suffering in the world is called *theodicy*. So far, this attempt has proven unsatisfactory in the judgment of the majority of philosophers and other scholars who have not already committed themselves to God as an act of faith.

The problem of evil remains the most powerful argument against God. But the argument collapses once any of the three omni- (3O) attributes are relaxed.

WHAT IS EVIL?

The argument from evil starts with the empirical fact that evil (bad stuff) exists in the world (a scientific statement) and shows that a god who is at the same time omnibenevolent, omnipotent, and omniscient—the 3O God—cannot exist.

We need to define evil before we can go much further. First we must confront what is called the *Euthyphro dilemma*.[3] Does God forbid us to do certain acts because they are evil, or is an act evil because God defines it as such?

Many of the same empirical facts about human behavior discussed in the previous chapter, which lead us to conclude that good exists independent of God, also apply to the case of evil. Evil does not seem to require the existence of God. As philosopher Kai Nielsen writes, "God or no God, torturing the innocents is vile. More generally, even if we can make nothing of the concept of God, we can readily come to appreciate . . . that, if anything is evil, inflicting or tolerating unnecessary and pointless suffering is evil, especially when something can be done about it."[4] An omnibenevolent, omnipotent, and omniscient God can do something about it.

Now, an easy escape from the argument from evil can be achieved by relaxing one or more of the three Os. For example, we can imagine a God who is not omniscient. Such a God would not always know when evil happened and so could not act to avoid it.

Similarly, a God who is not omnipotent may be unable to always stop evil. The latter possibility was the answer Rabbi Harold Kushner gave to the problem of evil in his best-selling book *When Bad Things Happen to Good People*.[5] Such a God can have a pleasant, human face, such as George Burns in the film *Oh, God!*[6] Burns, playing God, admits he isn't perfect. He says he would do things differently the next time he creates a universe. For one thing, he would not give the avocado such a large pit.

In a 1995 paper "Evil and Omnipotence,"[7] J. J. Mackie avers that adequate solutions to the problem of evil exist if you relax 3O, but he demonstrates the fallaciousness of several claimed solutions that retain the 3O God:

1. "Good cannot exist without evil" or "Evil is necessary as a counterpart to good."
2. "Evil is necessary as a means to good."

3. "The universe is better with some evil in it than it could be if there were no evil."
4. "Evil is due to human free will."

(Quotation marks in original.) Mackie shows that each of these solutions still implies a restriction on God's omnipotence. If God cannot create good without evil, that is a limit on his power. If God gives humans free will, then that is a restriction on his control of events.

Mackie gives a long rebuttal to argument 3 above. However, note that it is an example of the definitional problem mentioned in chapter 1. How do we define "better" so that a universe with more evil is better than another universe? We could just as well define the better universe as one with no evil.

One way that evil can coexist with omnipotence is if the evil is what philosophers call a "necessary truth." This is a statement that is true by virtue of its essential character. An example of such a statement is 2 + 2 is not equal to 5. This is true by virtue of the essential nature of numbers. Likewise, the statement that suffering is evil could be a necessary truth over which God has no power despite his omnipotence.[8]

Even so, this just means that God cannot simply define suffering as good. It does not prevent him from utilizing his power to eliminate or at least alleviate suffering.

Theologians have attempted to solve the problem of evil by pointing out that pain is a necessary part of life. Let's exempt such pain from our definition of evil and limit it to unnecessary pain. While pain warns us of disease and injury and prompts us to seek treatment, why must that pain persist, often unbearably, after treatment fails and we await death?

Another reason given for suffering is that it helps us to be compassionate. As theologian Richard Swinburne has put it, "If the world was without any natural evil and suffering we wouldn't have the opportunity . . . to show courage, patience and sympathy."[9]

But do we really need "natural evil" to encourage courage and sympathy? We can imagine a world in which the only pain was the necessary pain described above. A courageous act, such as giving up your life to save another's could be done in the absence of pain. Furthermore, many of the discomforts of life are not "natural evils" but necessities of growth—good in the benefits that accrue. For example, we can show sympathy to a child struggling through a difficult mathematics problem.

Does God really need so much pain and suffering to achieve his ends? Is there any conceivable good purpose behind so many children dying every day of starvation and disease? How are they helped by the rest of us becoming more sympathetic?

Yet another common theistic defense for the problem of evil is that God has given us the freedom to choose to commit evil. This may apply to the suffering that results from human acts; but, great as that may be, much unnecessary suffering is of natural rather than human origin. Examples include most diseases and natural disasters, such as the 2004 tsunami in Asia that killed hundreds of thousands of people. Indeed these are called "acts of God." And what is the purpose of the suffering of animals? Perhaps we can be sympathetic to that, but why is so much suffering necessary? And, what about the hundreds of millions of living things that died terrible deaths long before humans appeared on the scene?

In a department seminar in 2005 my University of Colorado colleague, philosopher Michael Huemer, provided a concise summary of current responses to the problem of evil and gave his personal analysis of why they all fail.[10] In the following, he is quoted exactly on these responses, but I give (in italics) my own short summary of reasons for their failure. Please excuse some repetitiveness here.

Summary of Attempts to Reconcile a 3O God with the Existence of Evil:

1. "Evil is a product of human free will. God gave us free will because free will is a very valuable thing. But he cannot both give us free will and prevent us from doing evil."

   Not all evil is the product of human free will, for example, natural disasters. If you redefine evil to include only human-caused ills, you still have to deal with the unnecessary suffering of natural disasters that are under God's control.

2. "Some amount of suffering is necessary for humans to develop important moral virtues. Some moral virtues can only exist in response to suffering or other bad things. Examples: courage, charity, strength of will."

   This could be accomplished with a whole lot less suffering than exists in the world.

3. "Good and evil exist only as contrasts to each other. Therefore, if evil were eliminated, good would automatically be eliminated as well."

   Good can exist independent of evil. Winning a race is good, but losing it is not evil. Buying a toy for your granddaughter is good, but not doing so is not evil when she already has a playroom full of toys.

4. "Slightly different from #3: If evil were eliminated, then we wouldn't *know* that everything was good, because we can only perceive things when there is contrast."

> *Even if we did not identify something as good, it still can be good. And it still can be good even if we have no experience of bad. My grandchildren know that having toys is good, although they have never had no toys and so have not had the opposing experience.*

5. "Perhaps God has a different conception of evil from ours. Maybe what we think of as evil is good."

> *We trust our own judgment on the evil of gratuitous suffering. No one can conceive of a reason God could have for allowing so much suffering. Why should we worship a God who allows acts that we regard as unspeakable? If God has a different conception of evil from ours, then so much the worse for God. He is then nothing more than an evil potentate. He might have power, but he has no moral authority and no one should worship him. "Good" and "evil" are our words and they name our concepts. It is confused thinking to suppose that some God's opinion would make any difference in our concepts.*

6. "Perhaps there is some underlying purpose served by all the evil in the world, but we humans are not smart enough to comprehend it. Have faith."

> *What could that possibly be? Again, why should we blindly accept acts that go against our very nature? Why would God give us a nature that finds his actions so reprehensible?*

7. "God is not responsible for evil. The Devil is."

> *The Judeo-Christian-Islamic God is stronger than the devil and so is still ultimately responsible.*

8. "If we simply weaken the definition of God, then the existence of God may be compatible with the existence of evil. Thus, for example, he might be unable to instantly eliminate all the evil."

> *While the Judeo-Christian-Islamic God described in scriptures is hardly benevolent, the faithful of these religions are far more likely to ignore unpleasant scriptural passages than abandon belief in a benevolent God.*

A huge philosophical and theological literature can be found on the problem of evil, which need not be summarized here. As throughout this book, the case will be presented as a scientific one. We have the undeniable empirical fact of considerable suffering in the world and have no reason to believe that the great bulk of that suffering is necessary. We have the hypothesis of a powerful God who is fully capable of alleviating all but the surely minimal suffering that is absolutely necessary. Many theologians argue that God has his own reasons for so much suffering, which then is, by definition, good. Our deepest instincts disagree and recognize unnecessary suffering as evil.

AN EVIL GOD?

We have seen that relaxing one of the Os, such as omniscience or omnipotence, can defeat the argument from evil against the existence of the 3O God. We can also relax omnibenevolence.

As should be clear to anyone who simply sits down and reads the Bible or Qur'an, the God described in these scriptures is hardly benevolent by normal human standards. Still, if you make Euthyphro's choice, then whatever God does is good by definition. In that case, for example, genocide and slavery are good.

In the previous chapter we saw that the Old Testament condones slavery. It also sanctions genocide: "Observe what I command you this day. Behold, I will drive out before you the Amorites, the Canaanites, the Hittites, the Per'izzites, the Hivites, and the Jeb'usites. Take heed to yourself, lest you make a covenant with the inhabitants of the land whither you go, lest it become a snare in the midst of you. You shall tear down their altars, and break their pillars, and cut down their Ashe'rim" (Exod. 34:11–13, Revised Standard Version).

Indeed, in the Old Testament, God admits he is the source of evil: "I form light and create darkness, I make weal and create woe, I am the LORD, who do all these things" (Isa. 45:7, Revised Standard Version).

The God of the Bible, if he exists, is not omnibenevolent by commonly accepted standards. At best, he is more like the dual God of Zoroastrianism, Manichaeism, and perhaps other religions—part good and part evil—or two separate but equal gods (or a pantheon of gods).

Interestingly, many Christians seem to regard Satan as a source of evil independent of God. Immediately after the September 11, 2001, tragedy, many (though by no means all) of the Christian clergy blamed it on the devil and not God.[11] This implies that either the devil is an equally powerful, autonomous separate God, which is no longer monotheism, or a part of God himself, which is no longer omnibenevolence.

If the theology of a dual god had survived, then we would have no problem of evil. Or, to put it better, evil would be a problem but we could blame it on God. However, monotheistic Christianity (albeit with the Trinity) had become the dominant religion in Europe in the fourth century when it gained the favor of Emperor Constantine (a pretty evil character in his own right).[12] Over the centuries, other variations were declared heretical and obliterated. In the doctrine that developed, Satan is still the creation of God but a fallen angel rather than a coequal deity. In that case, God is still the creator of evil.

We are once again confronted with the undeniable fact that our instincts about good and evil take precedence over supposed divine commands, when those commands offend both the common sense and the reason that has been cultivated over the centuries as humankind has gradually and incompletely evolved from brutish predecessors.

In the language of science, the empirical fact of unnecessary suffering in the world is inconsistent with a god who is omniscient, omnipotent, and omnibenevolent. Observations of human and animal suffering look just as they can be expected to look if there is no God.

NOTES

1. In a dialogue on religion with other scientists in 1999, quoted from "The Constitution Guarantees Freedom From Religion" an open letter to US vice presidential candidate Senator Joseph Lieberman, issued by the Freedom From Religion Foundation on August 28, 2000.

2. Michael Martin and Ricki Monnier, eds., *The Impossibility of God* (Amherst, NY: Prometheus Books, 2003), p. 59.

3. The dilemma is presented in Plato's *Euthyphro*, which is discussed in many philosophy books.

4. Kai Nielsen, *Ethics without God*, rev. ed. (Amherst, NY: Prometheus Books, 1990), p. 10.

5. Harold S. Kushner, *When Bad Things Happen to Good People* (New York: Avon Books, 1987).

6. Warner Bros., 1977.

7. J. J. Mackie, "Evil and Omnipotence," *Mind* 64 (1955): 200–12, reprinted in *The Impossibility of God*, ed. Martin and Monnier, pp. 61–105.

8. Erik J. Wielenberg, *Value and Virtue in a Godless Universe* (Cambridge: Cambridge University Press, 2005), p. 51.

9. Julian Baggini and Jeremy Stranghorn, *What Philosophers Think* (London: Continuum, 2003), p. 109.

10. Michael Huemer, "Some Failed Responses to the Problem of Evil," Talk at the Theology Forum, University of Colorado at Boulder, February 16, 2005.

11. See my discussion of religious reactions to the events of September 11, 2001, in Victor J. Stenger, *Has Science Found God? The Latest Results in the Search for Purpose in the Universe* (Amherst, NY: Prometheus Books, 2003), pp. 9–12.

12. Jonathan Kirsch, *God Against the Gods: The History of the War between Monotheism and Polytheism* (New York: Viking Compass, 2004).

Chapter 9

POSSIBLE AND
IMPOSSIBLE GODS

Why would a perfect God create a universe in which such huge amounts of suffering occur, when such suffering does not bring into existence any of the goods required to absorb the suffering and make the situation on balance a good one?

—Nicholas Everitt[1]

DISAGREEING WITH THE DATA

In this book I have applied the scientific process of hypothesizing models and testing those models against the empirical data to the question of God. Now, I am sure to hear the objection "science isn't everything." Of course it isn't. However, model building is not limited to science but is commonly carried out in everyday life, including religious activities. The brain does not

227

have the capacity to save the time, direction, and energy of each photon that hits the eyes. Instead it operates on a simplified picture of objects, be they rocks, trees, or people, assigning them general properties that do not encompass every detail.[2] Science merely objectifies the procedure, communicating by speech and writing among individuals who then attempt to reach agreement on what they all have seen and how best to represent their collective observations.

Religion carries out a similar process, although one in which agreement is generally asserted by authority rather than by consensus. From humanity's earliest days, gods have been imagined who possessed attributes that people could understand and to which they could relate. Gods and spirits took the form of the objects of experience: the sun, Earth, moon, animals, and humans. The gods of the ancient Egyptians had the form of animals. The gods of the ancient Greeks had the form of humans. The God of Judaism, Christianity, and Islam has the form of a powerful, autocratic, male king enthroned high above his subjects. Each seems to have developed from the culture of the day. If the process continued on to today, we would all worship cellular phones.

By dealing in terms of models of God that are based on human conceptions, we avoid the objection that the "true" God may lie beyond our limited cognitive capabilities. When I demonstrate that a particular God is rejected by the data, I am not proving that all conceivable gods do not exist. I am simply showing beyond a reasonable doubt that a God with the specific, hypothesized attributes does not exist. Belief aside, at the very minimum the fact that a specific God does not agree with the data is cause enough not to assume the existence of that God in the practices of everyday life.

The exact relationship between the elements of scientific models and whatever true reality lies out there is not of major concern. When scientists have a model that describes the data, that is consistent with other established models, and that can be put to practical use, what else do they need? The model works fine in not only describing the data but also in enabling practical applications. It makes absolutely no difference whether or not an electron is "real" when we apply the model of electrons flowing in an electronic circuit to design some high-tech device. Whatever the intrinsic reality, the model describes what we observe, and those observations are real enough.

Similarly, it does not matter from a practical standpoint whether the "real" God resembles any of the gods whose empirical consequences we have examined. People do not worship abstractions. They worship a God with qualities they can comprehend. Since we have shown that a God who answers prayers does not agree with the data, then a religious person is wasting her time praying for some favor of such a God. If praying worked, the effects would be objectively observed. They are not.

Let me then summarize the gods we have shown to disagree with the data. Again, an uppercase G will be used when the attributes apply specifically to the Judeo-Christian-Islamic God.

Gods Who Disagree with the Data

1. A God who is responsible for the complex structure of the world, especially living things, fails to agree with empirical fact that this structure can be understood to arise from simple natural processes and shows none of the expected signs of design. Indeed, the universe looks as it should look in the absence of design.

2. A God who has given humans immortal souls fails to agree with the empirical facts that human memories and personalities are determined by physical processes, that no nonphysical or extraphysical powers of the mind can be found, and that no evidence exists for an afterlife.

3. A God whose interactions with humans, including miraculous interventions, have been reported in scriptures is

contradicted by the lack of independent evidence that these miraculous events took place and the fact that physical evidence now convincingly demonstrates that some of the most important biblical narratives, such as the Exodus, never took place.

4. A God who miraculously and supernaturally created the universe fails to agree with the empirical fact that no violations of physical law were required to produce the universe, its laws, or its existence rather than nonexistence. It also fails to agree with established theories, based on empirical facts, which indicate that the universe began with maximum entropy and so bears no imprint of a creator.

5. A God who fine-tuned the laws and constants of physics for life, in particular human life, fails to agree with the fact that the universe is not congenial to human life, being tremendously wasteful of time, space, and matter from the human perspective. It also fails to agree with the fact that the universe is mostly composed of particles in random motion, with complex structures such as galaxies forming less than 4 percent of the mass and less than one particle out of a billion.

6. A God who communicates directly with humans by means of revelation fails to agree with the fact that no claimed revelation has ever been confirmed empirically, while many have been falsified. No claimed revelation contains information that could not have been already in the head of the person making the claim.

7. A God who is the source of morality and human values does not exist since the evidence shows that humans define morals and values for themselves. This is not "relative morality." Believers and nonbelievers alike agree on a common set of morals and values. Even the most devout decide for themselves what is good and what is bad. Nonbelievers behave no less morally than believers.

8. The existence of evil, in particular, gratuitous suffering, is logically inconsistent with an omniscient, omnibenevolent, omnipotent God (standard problem of evil).

WHAT IF?

The existence of the God worshiped by most Jews, Christians, and Muslims is not only missing from but also is contradicted by empirical data. However, it need not have turned out that way. Things might have been different, and this is important to understand as it justifies the use of science to address the God question and refutes the frequently heard statement that science can say nothing about God. If scientific observations had confirmed at least one model god, then even the most skeptical atheist would have to come around and admit that there might be some chance that God exists.

Consider the following hypothetical events that, had they occurred, would have favored the God hypothesis. The reader is invited to think of her own "might have been" scenarios that would force even the most dogmatic skeptic to reconsider his atheism.

Hypothetical Observations That Would Have Favored the God Hypothesis

1. Purely natural processes might have been proved incapable of producing the universe, as we know it, from nothing. For example, the measured mass density of the universe might not have turned out to be exactly what is required for the universe to have begun from a state of zero energy, which we assume is the energy of nothing. This would have implied that a miracle, the violation of energy conservation, was required to produce the universe.

2. Purely natural processes might have been proved incapable of producing the order of the universe. For example, suppose the universe were not expanding but rather turned out to be a firmament (as the Bible says it is). The second law of thermodynamics would require that the universe always had total entropy less than maximum in the past. Thus, if the universe had a beginning, that beginning would have to be one of order imposed from the outside. If the universe had no beginning but extended indefinitely into the past, then we still would need to account for the source of the ever-increasing order as we go back in time.

3. Purely natural processes might have proved incapable of producing the complex structure of the world. For example, the age of Earth might have turned out to be too short for evolution. Simple processes might not have been able to produce complex structure.

4. Evidence was found that falsified evolution. Fossils might have been found that were inexplicably out of sequence. Life-forms might not have all been based on the same genetic scheme. Transitional species might not have been observed.

5. Human memories and thoughts might have provided evidence that cannot be plausibly accounted for by known physical processes. Science might have confirmed exceptional powers of the mind that it could not plausibly explain physically. Science might have uncovered convincing evidence for an afterlife. For example, a person who has been declared dead by every means known to science might return to life with detailed stories of an afterlife containing information he could not possibly have known and is later verified as factual, such as the location of the nearest planet with life.

6. A nonphysical channel of communication might have

been empirically confirmed by revelations containing information that could not have been already in the head of the person reporting the revelation. For example, someone in a religious trance might learn the exact date of the end of the world, which then happens on schedule.

7. Physical and historical evidence might have been found for the miraculous events and the important narratives of the scriptures. For example, Roman records might have been found of an earthquake in Judea at the time of a crucifixion ordered by Pontius Pilate. Campsites might have been found in the Sinai Desert.

8. The void might have been found to be absolutely stable, requiring some action to bring something rather than nothing into existence.

9. The universe might have been found to be so congenial to human life that it must have been created with human life in mind. Humans might have been able to move from planet to planet, just as easily as they now move from continent to continent, and be able to survive on every planet without life support.

10. Natural events might follow some moral law, rather than morally neutral mathematical laws. For example, lightning might strike mostly wicked people; people who behave badly might fall sick more often; nuns would always survive plane crashes.

11. Believers might have had a higher moral sense than non-believers and other measurably superior qualities. For example, the jails might be filled with atheists while all believers live happy, prosperous, contented lives surrounded by loving families and pets.

But none of this has happened. The hypothesis of God is not confirmed by the data. Indeed that hypothesis is strongly contradicted by the data.

WHAT GODS REMAIN?

Now, a believer is certainly free to argue that "none of these Gods is my god." I have nowhere claimed that I can rule out every conceivable god, just those with the selected empirically detectable attributes. If a believer's god does not have any of those attributes, then I have no quarrel with her.

For example, we might imagine a god who created the universe but does not interfere with it or interact with its inhabitants in any way. The deist god of the Enlightenment (the "Creator" in the Declaration of Independence) created the universe with completely deterministic natural laws and thus has no need to ever step in. For this god, everything that happens is already written.

However, this type of deist god is probably ruled out by a fact drawn from most interpretations of quantum mechanics. Based on our best current knowledge, nature is not deterministic. The Heisenberg uncertainty principle of quantum mechanics implies that the motion of a particle cannot be predicted with absolute certainty, that much that happens in the universe is random. Furthermore, the latest developments in cosmology imply that the universe began in total chaos and so retains no memory of any creator. This still leaves open the possibility of a deist god who created the chaos and left everything else to chance. But, such a god has no observable effect and is functionally equivalent to nonexistent as far as humans are concerned.

In *Has Science Found God?* I mentioned those contemporary theologians who are making serious attempts to reconcile science and the supernatural.[3] I called them the "Premise Keepers," which perhaps was a bit cute but I did attempt to treat them with some sympathy. Their main concern is evolution, which they readily accept as well established. The problem they must deal with is the apparent accidental evolution of the human species. Some propose that God "poked his finger" in the historical process, so that humanity would appear. However, this is essentially intelligent design applied perhaps just once in evolution but applied nonetheless and contradictory to the essence of evolutionary theory.

Despite accepting the evolution of the human body, we saw that the Catholic Church insists that evolution does not apply to the mind.[4] While stating that he would change his Buddhist beliefs should science demonstrate any of them to be false, the Dalai Lama still insists that humans cannot be "reduced to nothing more than biological machines, the products of pure chance in the random combination of genes, with no purpose other than the biological imperative of reproduction."[5]

Some Premise Keepers are willing to accept the now apparent fact that humans are indeed biological machines that are products of chance. If you were to start the universe up again, we and every other species on Earth would not reappear in the same forms. Humanity is an accident. However, in the view of evolution theism, God can achieve his ends, whatever they are, by any of the countless pathways that become possible when no restrictions are placed on how matter may self-organize into complex systems.

Physicist Howard van Till imagines a "possibility space" of all potential life-forms. By means of random variations, God explores and discovers (in contrast to creates) novel life-forms that actualize his intentions in the course of time.[6] Einstein thought that God did not play dice with the universe, but the Premise Keepers say he does.

However, Phillip Johnson, the Christian lawyer who initiated the intelligent design movement, protests strongly that this is not Christianity but simply an updated deism, with God "exiled to that shadowy realm before the Big Bang" where he "must promise to do nothing that might cause trouble between theists and scientific naturalists."[7]

Obviously, neither the deist god nor the van Till god is the God of Jews, Christians, and Muslims. Their God plays a primary, moment-by-moment role in every phenomenon, from atomic collisions in the farthest galaxy down to the chemical reactions in each cell in each of the 10^{30} or so bacteria on Earth. And, of course, he reads every human thought. I have argued that such a God should have been detected by now, if not from casual observations, then surely from the precision data on every aspect of the world that have now been gathered by science. Like the chaos deity, a God with no observable effect is indistinguishable from one who is nonexistent. Certainly worshiping such a God serves no useful purpose.

While many laypeople have been led to believe that science has found evidence for God, this is simply not the case. If it were true, the news would have made simultaneous headlines in every newspaper in the world, using the "Second Coming" font reserved for stupendous events. Indeed, the Second Coming would provide the needed evidence. But it is now two thousand years overdue, Jesus having assured his disciples that he would return before they died.

As I have mentioned several times, there is no basis for the claim that science dogmatically refuses to accept the evidence for God, although some national scientific organizations, terrified of losing taxpayer support, have tried to distance science from religion. If scientific evidence for God turned up that passed the conventional tests applied to any extraordinary claim, then scientists in every field would be happily busy writing research grant proposals to study his nature. Instead most, even those who attend church on Sunday, go about their daily professional duties without ever bringing in God.

Serious theologians not committed by faith to their own dogma have gradually begun to accept the absence of objective evidence for God and have been forced to conclude if a god exists, he must purposely hide himself from us. I fully admit that possibility. God could simply work through natural processes and, indeed, may have reasons to hide himself from us. Let us see what kind of god that might be.

THE HIDDENNESS PROBLEM

In the fall of 2004, I attended a conference at the University of Colorado at Boulder on "The Hiddenness of God" sponsored by the Theology Forum of the Department of Philosophy. The attendees were mostly theologians, philosophers of religion, and other religious scholars, many from theology schools and mostly confirmed believers. They were seeking to find a rational explanation for what most seemed to readily accept as a fact: no empirical evidence for God exists.

One of the attendees was philosopher John L. Schellenberg, who opened the meeting with a presentation of what is called the *argument from hiddenness* for the nonexistence of God. He published this argument in a 1993 book, *Divine Hiddenness and Human Reason*.[8] Stated formally, the argument is as follows (as quoted from Schellenberg's handouts):

The Hiddenness Argument

1. If there is a perfectly loving God, all creatures capable of explicit and positively meaningful relationships with God who have not freely shut themselves off from God are in a position to participate in such a relationship that is, able to do so just by trying to.
2. No one can be in a position to participate in such relationship without believing that God exists.
3. If there is a perfectly loving God, all creatures capable of explicit and positively meaningful relationship with God who have not freely shut themselves off from God believe that God exists (from 1 and 2).
4. It is not the case that all creatures capable of explicit and positively meaningful relationship with God who have not freely shut themselves off from God believe that God exists: there is non-resistant nonbelief; "God is hidden."

5. It is not the case that there is a perfectly loving God (from 3 and 4).
6. If God exists, God is perfectly loving.
7. It is not the case that God exists (from 5 and 6).[9]

In short, a perfectly loving God would not deny knowledge of his existence to any human who is not resistant to that knowledge. The empirical fact that many humans are open to knowledge of God and still do not believe demonstrates that such a God does not exist.

This argument is similar to the *argument from nonbelief* of philosopher Theodore Drange, which Drange states as follows:

The Argument from Nonbelief

1. If God were to exist, then there would be no avoidable nontheism in the world.
2. But there is avoidable nontheism in the world.
3. Therefore, God does not exist.[10]

These arguments serve to answer the objection theists make to the argument from lack of evidence (see chap. 1) that God simply chooses to remain hidden from humanity. As Schellenberg puts it, "Why, we may ask, would God be hidden from us? Surely a morally perfect being—good, just, loving—would show himself more clearly. Hence the weakness of our evidence for God is not a sign that God is hidden; it is a revelation that God does not exist."[11] Conference participants agreed that the hiddenness argument is also connected to the problem of evil. For example, both concentrate on attributes that seem contradictory to the assumed moral character of God. I have only briefly discussed the problem of evil in this book (see chap. 8), since it is not a scientific argument and hardly original with me—although the existence of unnecessary suffering in the world is an empirical fact. However,

the problem of evil remains the strongest argument against a
beneficent God, one that theologians have grappled with for centuries without success.[12]

The hiddenness problem relates most directly to the scientific
arguments I have presented. If a theist attempts to refute my conclusions by claiming that God intentionally hides himself from us,
then that God cannot be the personal, perfect loving God of liberal
Christianity. However, there is another brand of Christian God.

THE HIDEOUS HIDDEN GOD
OF EVANGELICAL CHRISTIANITY

The believing theologians at the Boulder conference were all
Christians, and they provided a variety of responses. Jeff Cook, a
young graduate student at the University of Colorado, presented
a solution to the hiddenness problem that left the more evangel-
ically inclined Christians at the conference shaking their heads
vertically, while the rest of us shook our heads horizontally.

Cook gave some personal history of how being born again
turned his life around. His wife, sitting in the audience, was given
the chance to affirm their joint transforming experience.

Cook called his solution to the problem of divine hiddenness
the "Ecclesiastic solution." Let me use his own words, as presented in the conference handouts: "Christianity shows that one
of God's chief desires is to create a community of individuals that
are devoted to the good of one another, with God himself as the
chief participant. This community has many preconditions and
needs, and it may be the case that God's universal self-disclosure
would be less effective at creating and establishing the Kingdom
of God than a policy of selective self-disclosure."[13] In other
words, God does not wish to spend eternity with all human
souls, but only the chosen few who, by blind faith in the absence
of all evidence, accept a Jewish carpenter who may or may not

have lived two thousand years ago as their personal savior. Of course, this is hardly a new idea but was essentially the teaching of John Calvin (d. 1564).

To Christians of this persuasion, Mahatma Gandhi is burning in hell, along with the six million Jews killed by Hitler and the billions of others who have died without accepting Jesus.

Those Catholics and evangelical Christians who hold this view clearly do not believe in a perfectly loving God. Their God dooms everyone else but them to eternal fire. Muslims, too, insist that theirs is the only way to salvation. And while the range of belief in modern Judaism is enormous, including many Jews who are athe-ists but still practice their religion out of respect for their heritage, a few extreme Jews still regard themselves as the chosen people of God. If anyone promoted such views in any area outside a reli-gious context, he would be taken in for psychiatric evaluation.

Philosopher Evan Fales has given another explanation for the hiddenness of God: "Some apologists tell us that God remains hidden from us so as not to coerce our worship. But God is not hiding out of solicitude for our freedom. We have not forgotten Job: therefore we understand that God is hiding out of cowardice. God is in hiding because He has too much to hide. We do not seek burning bushes or a pillar of smoke. No—we wish to see God. Can God stand before us? Can God see the face of suffering humanity—and live?"[14]

The existence of the Catholic, evangelical, extreme Muslim, extreme Judaic God who hides himself from all but a selected elite cannot be totally ruled out. All I can say is that we have not one iota of evidence that he exists and, if he does exist, I person-ally want nothing to do with him. This is a possible god, but a hideous one.

NOTES

1. Nicholas Everitt, *The Non-Existence of God* (London, New York: Routledge, 2004), p. 236.

2. Paul Bloom, *Descartes' Baby: How the Science of Child Development Explains What Makes Us Human* (New York: Basic Books, 2004).

3. Victor J. Stenger, *Has Science Found God? The Latest Results in the Search for Purpose in the Universe* (Amherst, NY: Prometheus Books, 2003), chap. 11.

4. Pope John Paul II, Address to the Academy of Sciences, October 28, 1986, *L'Osservatore Romano*, English ed., November 24, 1986, p. 22.

5. Dalai Lama, *The Universe in a Single Atom: The Convergence of Science and Spirituality* (New York: Random House, 2005).

6. Howard van Till, in Phillip E. Johnson and Howard van Till, "God and Evolution: An Exchange," *First Things* 34 (1993): 32–41.

7. Ibid.

8. John L. Schellenberg, *Divine Hiddenness and Human Reason* (Ithaca, NY: Cornell University Press, 1993).

9. Ibid., "The Problem of Hiddenness and the Problem of Evil," Presented to the Conference on "The Hiddenness of God," Theology Forum, University of Colorado at Boulder, October 21–23, 2004.

10. Theodore M. Drange, *Nonbelief and Evil: Two Arguments for the Nonexistence of God* (Amherst, NY: Prometheus Books, 1998), p. 23.

11. Schellenberg, *Divine Hiddenness and Human Reason*, p. 1.

12. J. J. Mackie, "Evil and Omnipotence," *Mind* 64 (1955): 200–12; Keith Parsons, *God and the Burden of Proof: Platinga, Swinburne, and the Analytical Defense of Theism* (Amherst, NY: Prometheus Books, 1989); Drange, *Nonbelief and Evil*.

13. Jeff Cook, "The Problem of Divine Hiddenness," Presented to the Conference on "The Hiddenness of God," Theology Forum, University of Colorado at Boulder, October 21–23, 2004.

14. Evan Fales, "Despair, Optimism, and Rebellion," http://www.infidels.org/library/modern/evan_fales/despair.html (accessed July 6, 2005).

Chapter 10

LIVING IN THE GODLESS UNIVERSE

Live joyfully with the wife whom thou lovest all the days of the life of thy vanity, which he hath given thee under the sun, all the days of thy vanity: for that is *thy portion in* this *life, and in thy labor which thou takest under the sun. Whatsoever thy hand findeth to do, do it with thy might; for* there is *no work, nor device, nor knowledge, nor wisdom, in the grave whither thou goest.*
—Ecclesiastes 9: 9–10 (King James Version)

IS RELIGION USEFUL?

Archaeology testifies that religion was a major component of human life for thousands of years before civilization began. And, of course, civilization did not put an end to religion but molded it into more sophisticated forms. The God of

243

244 GOD: THE FAILED HYPOTHESIS

Judaism, Christianity, and Islam arose in parallel with the city-state and may have been created to justify the relation of an all-powerful king to his subjects.

Voltaire (d. 1778) said, "If God did not exist, it would be necessary to invent him."[1] Of course, the French philosopher and satirist was being his usual cynical self, but a common view is that religion is a necessary component of human life. The reason most frequently given is that, without religion, everyone would behave immorally and society would be wracked by wars and all kinds of other evils. However, despite the widespread influence of religion, some humans continue to behave immorally and some morally, with no particular correlation to faith being evident. And, society continues to be wracked by wars and all kinds of other evils. If we need yet one more example of a failed model, this is it.

Yet most people believe, and libraries are full theories as to why they do despite all the evidence to the contrary.[2] Justin L. Barrett asks directly, *Why Would Anyone Believe in God?* and credits such belief to the types of mental tools we all carry around in our brains.[3] We hear other speculations that religious belief is built into our brains, with a special "God module"[4] perhaps codified by a "God gene"[5]—all the result of natural selection.

Psychologist Paul Bloom refers to recent research by himself and others indicating that the human brain has evolved two separate "programs" for analyzing the data from the senses.[6] One program deals with physical objects and the other with social relationships. Bloom suggests that this has led to a natural, built-in tendency to separate the world of matter from the world of mind and to believe in the survival of personality after death. As anthropologist Pascal Boyer has suggested, this also leads to a strong tendency to see purpose and design even when they are not there.[7] Anthropologist Stewart Guthrie calls this hypersensitivity to signs of agency, seeing intention where there is only accident or artifice, the clothes have no emperor.[8] Columnist

Nicholas Kristof of the *New York Times* finds this a "cosmic joke," that "humans have gradually evolved to leave many of us doubting evolution."[9]

If religion is a naturally evolved trait, then we have yet another argument against the existence of God. As always, the apologist can counter with the claim that God could still be behind it all. However, he can provide no evidence to support that hypothesis or any reason for introducing it. Once again, God is simply not needed any more than Bigfoot, the Abominable Snowman, and the Loch Ness Monster.

The issue of a God module in the brain remains controversial, and we will have to wait and see. The timescale would seem too short for biological evolution of such a major nature. On the other hand, the timescale is long enough for cultural evolution. We can still consider the implications of the proposition that religion had survival value, whether or not this resulted in the humans evolving some genetic propensity for religion that became built into their genes. Religion may be a cultural idea that evolved by natural selection because it provided a survival benefit, sort of the way the idea of traffic lights evolved.

THE NEGATIVE IMPACT OF RELIGION ON SOCIETY

In chapter 7, we saw how our notions of morals and values may have evolved naturally, their precursors being seen in animal behavior. There I argued that we possess innate concepts of what is good and what is bad that do not derive from a divine source and are, indeed, contradicted by the scriptures that are supposed to have a divine source.

We saw that the empirical evidence does not support the widespread assertion that religion is especially beneficial to society as a whole. Of course, it has always proved extremely beneficial to those in power—helping them to retain that power—from prehistoric times to the latest presidential election. But it is not clear how society is any better off than it would have been had the idea of gods and spirits never evolved.

Morality and religion may have evolved together. We can easily imagine, and history seems to confirm, that religion was the means by which good behavior—"good" usually being defined by whomever was in power at the time—was enforced. Even in modern times we see the remnants of this unholy alliance, with world leaders asserting divine authority for their actions and people still falling for it. By claiming divine authority, politicians are able to promote policies of dubious value that the public might otherwise find unacceptable. Journalist Chris Mooney has provided many disgraceful recent examples of this in his book *The Republican War on Science*.[10]

In February 2003 President of the United States George W. Bush told Australian Prime Minister John Howard that liberating the people of Iraq would not be a gift provided by the United States but, rather, "God's gift to every human being in the world."[11] In November 2004 Bush was reelected by a majority that included many who sincerely believed the president was carrying out God's work.

Theists in the United States continue to insist, contrary to the historical facts, that God is the foundation of our political system and that we and our political leaders must all abide by their particular interpretations of God's will. As Father Frank Pavone of the antiabortion organization Priests for Life told the 2000 Republican National Convention, "The Church does not dictate the policies of the nation. The Church proclaims the truth of God to which all these [public] policies must conform."[12]

A far more powerful figure who holds this view and applies it with a vengeance in his decisions is US Supreme Court Associate Justice Antonin Scalia. He quotes St. Paul:

Let every soul be subject unto the higher powers. For there is no power but of God: the powers that be are ordained of God. Whosoever therefore resisteth the power, resisteth the ordinance of God: and they that resist shall receive to themselves damnation. For rulers are not a terror to good works, but to the evil. Wilt thou then not be afraid of the power? Do that which is good, and thou shalt have praise of the same: for he is the minister of God to thee for good. But if thou do that which is evil, be afraid; for he beareth not the sword in vain: for he is the minister of God, a revenger to execute wrath upon him that doeth evil. Wherefore ye must needs be subject, not only for wrath, but also for conscience sake. (Romans 13:1–5, King James Version)

Scalia has declared, "Government—however you want to limit that concept—derives its moral authority from God."[13] He and Father Pavone apparently would have the United States abandon the Declaration of Independence: "We hold these truths to be self-evident, that all men are created equal, that they are endowed by their Creator with certain unalienable Rights, that among these are Life, Liberty and the pursuit of Happiness.—That to secure these rights, Governments are instituted among Men, deriving their just powers from the consent of the governed . . ." Although American Christians have been led to believe that the "Creator" mentioned here is their God, Thomas Jefferson, who wrote these words, was not a Christian but a deist. But my point here is that Scalia and Pavone reject the authority of the governed in favor of the authority of God, as they interpret his authority for us, of course.

According to Scalia, who President Bush called his model for Supreme Court appointments, governments do not derive "their just powers from the consent of the governed." Rather, Scalia tells us, "It [government] is the 'minister of God' with powers to 'revenge' and to 'execute wrath', including even wrath by the sword [which is unmistakably a reference to the death penalty]."[14] In March 2005 the United States became the last country in the world to abolish the death penalty for offenders who were under eighteen when they committed murder. Scalia vigorously dissented from the Supreme Court decision.

Most Americans view the Constitution as a "living document" that evolves as society evolves. Scalia calls this a "fallacy." For him, the text is fixed in meaning what it always meant. If slavery, which was not forbidden in the Constitution, still existed today, Scalia would probably rule against its abolition. If women could not vote, Scalia would do his best to see that they never did. No doubt he would use the Bible to justify those opinions.

Justice Scalia's thinking exemplifies all that is wrong with religion and why it is so inimical to human progress. God rules over a physical and social firmament that must remain unchanged, because change implies imperfection in his original creation.

I hope I have made it clear in this book that, while I wish people were less gullible, less willing to believe in the most preposterous supernatural notions, I still have a high regard for the basic decency of most human beings. Many people are good. But they are not good because of religion. They are good despite religion.

Nineteen Muslims would not have wreaked the havoc of September 11, 2001, destroying themselves along with three thousand others, had they not been believers. I need not detail all the killing in the name of God that has gone on throughout the ages.[15] At the time of this writing we have religious conflicts in a half-dozen places around the world.[16] In his book *Is Religion Killing Us?* Jack Nelson-Pallmeyer traces the biblical and Qur'anic sources of violence. He concludes, "Violence is widely embraced because it is embedded and 'sanctified' in sacred texts and because its use seems logical in a violent world."[17]

Religion at least partially accounts for the large cultural differences and mistrust that divide racially similar groups, like Israelis and Palestinians or Indians and Pakistanis, who might otherwise live together in harmony or even as a single people.

Not every war in history has been over religion, but religion has done little to ameliorate the conditions that led to war in those cases. We just have to look back half a century and witness the role played by the Catholic Church in aiding Nazi Germany.[18] For example, the German Church opened its genealogical records to the Third Reich so that a person's Jewish ancestry could be traced. Not a single German Catholic, including Adolf Hitler, was excommunicated for committing crimes against humanity.[19] And Hitler often claimed he was serving God. In *Mein Kampf* he says, "Hence today I believe that I am acting in accordance with the will of the Almighty Creator: *by defending myself against the Jew, I am fighting for the work of the Lord.*"[20] However, I must hasten to add that many Catholic leaders outside Germany did speak out against the Nazis and some, such as the Dutch archbishop, were retaliated against.

Now, you might ask, what about all the undeniable good that is done by religious charitable institutions—helping the poor and caring for the afflicted? Although the many selfless and dedicated people who do charitable work will tell you that they are motivated by their love of God, it is not really clear that God has that much to do with it. Perhaps these people are simply innately charitable and would have done the same in the absence of religious motives. The empirical fact is that people with no religion are not noticeably less charitable than those with religion.

Much of the time and money spent by Christian charities, including that now provided by federal and state governments in the United States as part of "faith-based initiatives," goes to proselytizing rather than solving the problems they were set up to solve. This money would be put to better use in providing services other than worship services. Certainly no evidence exists that so-called faith-based charities do any better than secular ones. Indeed, there is mounting evidence that some do worse.

For example, in 1996, the then Texas governor George W. Bush saw to it that state agencies eliminated inspection requirements of religious charities. In five years, the rate of confirmed abuse and neglect at religious facilities rose by a factor of twenty-five compared to state-licensed facilities. In another example of misuse, a Texas state district court found that a jobs training program unconstitutionally used $8,000 of state money to buy Bibles and spent most of the time on Bible study while providing no secular alternatives.[21] For a survey of the negative social impact of religious extremism in the United States, see the book of essays edited by Kimberly Blaker.[22]

Now, you might say this has nothing to do with the existence or nonexistence of God. However, the concept of a beneficent, loving God held by most people would reasonably be expected to lead to a better world when God is widely worshiped. Well, God is widely worshiped and we do not have a better world because of it. On the contrary, the world seems worse off as the result of faith. The certainty and exclusiveness of the major monotheisms make tolerance of differences very difficult to achieve, and these differences are the major source of conflict.[23]

In stark contrast to almost all other religious leaders, the Dalai Lama has tried to keep Tibetan Buddhism in tune with the modern world. He has often made it clear that whenever a Buddhist teaching disagreed with science, then he would attempt to change the teaching. However, as I have already noted, the Dalai Lama still seems to believe in a duality of mind and body not supported by science.

Not that Buddhists have avoided committing their own atrocities (condemned by the Dalai Lama, to be sure), as the recent history of Sri Lanka demonstrates.

MEANING

Finally, we need to deal with the personal aspects of religion that may be the most important for most people. In this section we discuss the common claim that life is meaningless if God does not exist.[24] In the next section we will consider the widespread belief that religion provides comfort and inspiration.

Christian apologist William Lane Craig has spoken of "the absurdity of life without God." According to science, the human race is ultimately doomed as the universe plunges toward inevitable extinction. Without God, without immortality, Craig tells us, "The life we live is without ultimate significance, ultimate value, ultimate purpose."[25]

Philosopher Erik Wielenberg tells of a gym teacher who would calm things down when tempers flared during a heated ball game by saying, "Ten years from now, will any of you care who won this game?" Wielenberg recalls thinking that a reasonable response would be, "Does it really matter *now* whether any of us will care in ten years?"[26] He quotes philosopher Thomas Nagel in the same vein, "It does not matter now that in a million years nothing we do now will matter."[27]

In other words, what matters now is what happens now. The September 11, 2001, hijackers were guided by some imagined ultimate purpose and so did not care what happened to them when they flew airplanes into buildings. We (mostly) all agree how sick that was. We can take comfort in that highly probable fact that they did not wake up in paradise.

Surely we can find present meaning in our lives that does not depend on our immortality, especially since our immortality is not likely to happen. Independent of immortality, many people think that life is pointless unless they fit into some grand, cosmic scheme. They imagine that meaning can only be assigned externally, by some outside, higher authority.

But, why can't we find meaning internally? Why must meaning be handed down from above? Over the ages, philosophers have offered many suggestions on how to live rewarding lives. In his *Nicomachean Ethics*, Aristotle offered three ways that humans might live contentedly: a life devoted to the pursuit of bodily pleasure; a life devoted to political activity; and a life devoted to contemplation.[28] He decided that the life of contemplation was best, since that most closely matches the activity of the gods. I suppose he wasn't thinking of the gods in Homer's *Iliad*.

Many theists will claim that, without God, humans would seek only bodily pleasure and other selfish interests. But that is not the nature of a social animal. We seek pleasure in the society of others and we empathize with others suffering. With the evolution of civilization, we have an enormous range of wonderful and important activities in which we can participate. I got my curiosity from the same place as cats, but I've been able to pursue mine into the deepest questions about the nature of the universe with the help of multimillion-dollar instruments and thousands of other scientists. Far from providing us meaningful goals, religions prescribe tribal values: amity for our tribe; enmity for other tribes; mind-closing faith; abject worship of authority.

God is not necessary for someone to find fulfillment in contemplation or social activity. Ethical philosopher Peter Singer emphasizes that "[we] can live a meaningful life by working toward goals that are objectively worthwhile."[29] One of the ways he suggests is quite simple, namely, to work to reduce avoidable suffering. This, he claims, is an objectively worthwhile goal that can provide inner meaning and, furthermore, can be done whether or not God exists.

Similarly, philosopher Kai Nielsen has remarked, "A man who says, 'If God is dead, nothing matters,' is a spoilt child who has never looked at his fellow man with compassion."[30]

COMFORT AND INSPIRATION

Many find comfort and inspiration in the notion that they are not alone in the universe, that they are a special part of the cosmos with a loving father looking down on them and providing them with an eternal life. During their mortal lives, many also claim that religion inspires them to do greater things, to go beyond the bounds of their material existences.

The idea of life after death probably came about when our primitive ancestors evolved the cognitive ability to not only realize that they will someday die, but also to ask whether death is final or that something still lay beyond the grave. The latter possibility would have been strongly suggested by the fact that a dead person was still "alive" in thoughts and dreams. Those thoughts and dreams were ephemeral, so the notion arose that some "spirit" carried on after the material body ceased to move and began to decay.

In chapter 3 we traced the development of the soul to the place where it exists today as little more than a word used to represent someone's "personhood," encompassing the qualities such as love and kindness that identify a person as something more than a mechanical automaton. It now seems almost certain that those qualities are not the product of some immaterial substance or spirit but arise through the natural operations performed by a highly complex but still purely material brain. That brain dies when we die, but our memories and thoughts carry on in the brains of others.

Unfortunately, science cannot confirm the Christian-Islamic promise that we one day will be reunited with departed loved ones and live eternally in the bosom of our creator.[31] The rational prospect of life after death is close to nil. But, at least, science *can* assure us that the many who happened to choose the wrong God will not be tortured through all eternity—that those millions who lived and died before the jealous God was invented will rest in peace. As an atheist T-shirt says, "Smile. There is no hell."

Science can help us to live a better life with the years we have. No doubt most of humanity today enjoys longer lives in greater comfort and pleasure as the direct result of scientific advances (such as evolution) than it would in the absence of those advances—especially if humanity had relied solely on religious teachings. If science has brought with it new problems, such as overpopulation, pollution, and the threat of nuclear holocaust, few people suggest we do away with science to avoid those consequences. Hopefully they can be avoided with the help of science and wise political actions.

Additionally, by ridding the world of superstition, science helps us live in less fear of the unknown. Humans no longer cower in the back of a cave during an electrical storm—and they know enough to get off the golf course. People are no longer burned at the stake when accused of heresy or witchcraft. By ridding the world of God, science helps us to control our own lives rather than submitting them to the arbitrary authority of priests and kings who justify their acts by divine will.

I do not deny that religion has inspired great art and music, which does much to enrich our lives. I personally have spent many happy hours viewing religious art in the great museums of the world and listening to sacred music in concert halls and recordings. I cannot think of anything more beautiful or more touching (or amazing) than Michelangelo's *Pieta* in St. Peter's in Rome. In my youth I thrilled at singing Bach's "Magnificat," Handel's *Messiah*, and Brahms's *Requiem* as a member of a church choir.

Many religious stories appeal to us as poetry and plays do. They are parables speaking to the human condition. Their value has nothing to do with the supernatural or whether they are true. Most have existed in many forms, both religious and secular: Moses in the bulrushes; the ugly duckling; Luke Skywalker on Tatooine. Don't all young people feel that they may have greatness in them—and why shouldn't they? David and Goliath; Jack the Giant Killer; Odysseus and the Cyclops. We all need the courage to never give up, to call on our ingenuity and initiative against the giants we face.

Beauty and inspiration can arise from secular sources. Certainly much great art and literature is secular in nature. Religion hardly comes up in Shakespeare, the greatest poet of the English language. Often romantic love is the inspiration for great poetry, as when Romeo calls up to Juliet from her garden at sunrise,

> Arise, fair sun, and kill the envious moon,
> That is already sick and pale with grief,
> That thou her maid art far more fair than she.[32]

Many people think of science as cold and impersonal. Scientists have tried to counter that by pointing to the beauty and majesty of nature and the great pleasure and inspiration that science brings to its practitioners. In his 1980 hit public television series, *Cosmos*, astronomer Carl Sagan extolled the grandeur of the universe, life, and the human brain. In his book *Pale Blue Dot*, Sagan asks, "How is it that hardly any major religion has looked at science and concluded, 'This is better than we thought! The Universe is much bigger than our prophets said, grander, more subtle, more elegant'? Instead they say, 'No, no, no! My god is a little god, and I want him to stay that way.' A religion, old or new, that stressed the magnificence of the Universe as revealed by modern science might be able to draw forth reserves of reverence and awe hardly tapped by the conventional faiths."[33]

In his 1998 book, *Unweaving the Rainbow*, Richard Dawkins carried on in the Sagan tradition: "The feeling of awed wonder that science can give us is one of the highest experiences of which the human psyche is capable. It is a deep aesthetic passion to rank with the finest that music and poetry can deliver. It is truly one of the things that makes life worth living and it does so, if anything, more effectively if it convinces us that the time we have for living it is fragile."[34]

Dawkins takes his title from a poem by John Keats:

> Philosophy will clip an Angel's wings,
> Conquer all mysteries by rule and line,
> Empty the haunted air, and gnomed mine—
> Unweave a rainbow . . .[35]

Keats felt that Newton had destroyed the poetry of the rainbow by reducing it to the prismatic colors.[36] Dawkins disagrees, pointing out how the unweaving of the rainbow—the separation of its components into different wavelengths—adds to rather than detracts from its beauty and poetry. The threads of the rainbow have been rewoven into the beautiful tapestry of modern physical and biological science. From the spectral threads of visible light, a model of the atomic structure of matter has been woven. From the spectral threads of x-rays reflected off the atoms of biological matter, a model of the structure of the key to life, DNA, has been woven. From the spectral threads of light from stars and galaxies, and more recently that of the cosmic microwave background radiation, a model of the structure of the universe has been woven.

Dawkins expresses the fulfillment of being a scientist:

After sleeping through a hundred million centuries we have finally opened our eyes on a sumptuous planet, sparkling with color and bountiful with life. Within decades we must close our eyes again. Isn't it a noble, an enlightened way of spending our brief time in the sun, to work at understanding the universe and how we have come to wake up in it? This is how I answer when I am asked—as I am surprisingly often—why I bother to get up in the mornings. To put it the other way round, isn't it sad to go to your grave without ever wondering why you were born? Who, with such a thought, would not spring from bed, eager to resume discovering the world and rejoicing to be part of it?[37]

Dawkins wishes he had written the following quatrain by William Blake, saying the meaning and inspiration would have been very different from that of the mystical Blake,

> To see a world in a grain of sand,
> And a heaven in a wild flower,
> Hold infinity in the palm of your hand,
> And eternity in an hour.[38]

Of course, most people value the benefits of science. Every-where you go these days you see people talking on their mobile phones. They don't have to pass a course in the theory of electro-magnetic waves before using them. But they also miss the exqui-site pleasure of writing down the four beautiful equations of elec-tromagnetism, called *Maxwell's equations*, and deriving from them other equations that describe the propagation of electromagnetic waves in a vacuum that move at exactly the speed of light.

Nevertheless, our mobile phone user can still obtain ample inspiration and pleasure in art, music, literature, and the more mundane but equally important events of everyday life—family, work, and recreation. At least science helps make this possible by freeing humans from the need to spend all their time on simple survival. Unfortunately we still live in a world where this freedom is not yet enjoyed by all.

So, even though science is a valuable tool available to most of humanity, only a tiny few find it a source of inspiration and even fewer a source of comfort. Religion, on the other hand, is sup-posed to provide comfort for all. However, religious comfort is not all that it is cracked up to be. In a recent study, psychologists found that highly religious Protestants exhibit more symptoms of obsessive-compulsive disorder than the less religious or nonreli-gious.[39] The promise of life after death carries with it the dread that the afterworld will be spent elsewhere than in the bosom of God. Everyone is a sinner, and even the most cloistered nun lives with the nagging worry that she might not be forgiven for that occasional impious thought that slips into her head between endless recitations of the Hail Mary. Likewise, the believer in rein-carnation might sometimes worry about living his next life as a rodent. The Muslim suicide bomber has been led to believe that he is guaranteed paradise by his murderous action. On the other hand, the atheist has the comfort of no fears for an afterlife and lacks any compulsion to blow himself up.

No doubt a temporary feeling of peace of mind can be

achieved during prayer or meditation. This results from an emptying of the mind of other thoughts, especially thoughts of self. Of all the world's religions, Buddhism provides the clearest understanding of the process, although every indication is that the mechanism is purely physical.[40] Enlightenment can only be obtained when the individual is able to eliminate all the desires of self. Nirvana is not heaven. Nirvana is nothingness.

However, I am not quite ready for nothingness. I am willing to trade nirvana for the joy and anguish of life for at least a few more years.

NOTES

1. Voltaire, *Candide, ou l'Optimisme*, first published in 1759.

2. See, for example, Daniel Dennett, *Breaking the Spell: Religion as a Natural Phenomenon* (New York: Viking Penguin, 2006).

3. Justin L. Barrett, *Why Would Anyone Believe in God?* (Walnut Creek, CA: AltaMira Press, 2004).

4. V. S. Ramachandran, "God and the Temporal Lobes of the Brain," Talk at the conference Human Selves and Transcendental Experiences: A Dialogue of Science and Religion, San Diego, California, January 31, 1998; Matthew Alper, *The "God" Part of the Brain: A Scientific Interpretation of Human Spirituality and God* (Brooklyn, NY: Rogue Press, 2001); Andrew Newberg and Eugene d'Aquili, *Why God Won't Go Away* (New York: Ballantine Books, 2001); Pascal Boyer, *Religion Explained: The Evolutionary Origin of Religious Thought* (New York: Basic Books, 2001); Donald M. Broom, *The Evolution of Morality and Religion* (Cambridge: Cambridge University Press, 2003).

5. Dean H. Hamer, *The God Gene: How Faith Is Hardwired into Our Genes* (New York: Doubleday, 2004).

6. Paul Bloom, *Descartes' Baby: How the Science of Child Development Explains What Makes Us Human* (New York: Basic Books. 2004); "Is God an Accident?" *Atlantic* 296, no. 5 (December 2005): 105–12.

7. Boyer, *Religion Explained.*

8. Stewart Elliott Guthrie, *Faces in the Clouds: A New Theory of Religion* (New York, Oxford: Oxford University Press, 1993).

9. Nicholas D. Kristof, "God and Evolution," op-ed, *New York Times*, February 12, 2005, p. 17.

10. Chris Mooney, *The Republican War on Science* (New York: Perseus Books Group, 2005).

11. Peter Singer, *The President of God and Evil: The Ethics of George W. Bush* (New York: Dutton, 2004), p. 208.

12. As quoted in Kimberly Blaker, ed., *The Fundamentals of Extremism: The Christian Right in America* (New Boston, MI: New Boston Books, 2003), p. 13.

13. Antonin Scalia, "God's Justice and Ours," *First Things* 123 (May 2002): 17–21. Online at http://www.firstthings.com/ftissues/ft0205/articles/scalia.html (accessed March 15, 2005).

14. Ibid.

15. See, for example, James A. Haught, *Holy Horrors: An Illustrated History of Religious Murder and Madness* (Amherst, NY: Prometheus Books, 1990).

16. Sam Harris, *The End of Faith: Religion, Terror, and the Future of Reason* (New York: Norton, 2004), p. 26.

17. Jack Nelson-Pallmeyer, *Is Religion Killing Us? Violence in the Bible and the Quran* (Harrisburg, PA: Trinity Press International, 2003), p. 146.

18. Gregory S. Paul, "The Great Scandal: Christianity's Role in the Rise of the Nazis," *Free Inquiry* 23, no. 4 (October/November 2003): 20–29; 24, no. 1 (December 2003/January 2004): 28–34.

19. Ibid, pp. 103–104.

20. Adolf Hitler, *Mein Kampf*, vol. 1, chap. 2.

21. Don Monkerud, "Faith No More," *In These Times* 27, no. 9 (March 10, 2003). Online at http://inthesetimes.com/comments.php ?id=105_0_2_0_C (accessed February 19, 2005).

22. Blaker, *The Fundamentals of Extremism*.

23. For further discussion of this point, see Harris, *The End of Faith*; Richard Dawkins, *The God Delusion* (Boston, New York: Houghton Mifflin, 2006); Sam Harris, *Letter to a Christian Nation* (New York: Alfred A. Knopf, 2006).

24. In this section I have relied heavily on Erik J. Wielenberg, *Value and Virtue in a Godless Universe* (Cambridge, New York: Cambridge University Press, 2005).

25. William Lane Craig, "The Absurdity of Life without God." Online at http://www.hisdefense.org/audio/wc_audio.html (accessed March 9, 2004).

26. Wielenberg, *Value and Virtue in a Godless Universe*, p. 30.

27. Thomas Nagel, *Mortal Questions* (Cambridge: Cambridge University Press, 1979), p. 11.

28. Aristotle, *Nicomachean Ethics*, trans. Martin Ostwald (Englewood Cliffs, NJ: Prentice-Hall, 1962), p. 8. See also Wielenberg, *Value and Virtue in a Godless Universe*, pp. 24–25.

29. Singer, *How Are We to Live? Ethics in an Age of Self-Interest* (Amherst, NY: Prometheus Books, 1995), p. 195.

30. Kai Nielsen, *Ethics without God*, rev. ed. (Amherst, NY: Prometheus Books, 1990), pp. 227–28.

31. Judaism does not have the strong emphasis, indeed obsession, with eternal life found in Christianity and Islam.

32. Shakespeare, *Romeo and Juliet.*

33. Carl Sagan, *Pale Blue Dot* (London: Headliner, 1995).

34. Richard Dawkins, *Unweaving the Rainbow: Science, Delusion and the Appetite for Wonder* (Boston, New York: Houghton Mifflin Co., 1998), p. x.

35. John Keats, "Lamia" (1820).

36. Dawkins, *Unweaving the Rainbow*, p. 39.

37. Ibid, p. 6.

38. William Blake, "Auguries of Innocence" (c. 1803).

39. Jonathan S. Abramowitz, Brett J. Deacon, Carol M. Woods, and David F. Tolin, "Association between Protestant Religiosity and Obsessive–Compulsive Symptoms and Cognitions," *Depression and Anxiety* 20 (2004): 70–76.

40. Antonie Lutz, Lawrence L. Greischar, Nancy B. Rawlings, Mathew Ricard, and Richard J. Davidson, "Long-term Meditators Self-induce High-amplitude Gamma Synchrony during Mental Practice," *Proceedings of the National Academy of Sciences* 101, no. 46 (2004): 16369–73.

BIBLIOGRAPHY

Abramowitz, Jonathan S., Brett J. Deacon, Carol M. Woods, and David F. Tolin. "Association between Protestant Religiosity and Obsessive–Compulsive Symptoms and Cognitions." *Depression and Anxiety* 20 (2004): 70–76.

Acocella, Joan. "Holy Smoke; What Were the Crusades Really About?" *New Yorker*, December 13, 2004.

Adami, Christoph. *Introduction to Artificial Life*. New York: Springer, 1998.

Adami, Christoph, Charles Ofria, and Travis C. Collier. "Evolution of Biological Complexity." *Proceedings of the National Academy of Sciences USA* 97 (2000): 4463–68.

Aguire, Anthony. "The Cold Big-Bang Cosmology as a Counter-example to Several Anthropic Arguments." *Physical Review* D64 (2001): 083508.

Aine, C. J. "A Conceptual Overview and Critique of Functional Neuro-Imaging Techniques in Humans: I. MRI/fMRI and PET." *Critical Reviews in Neurobiology* 9, nos. 2–3 (1995): 229–309.

Alexander, Richard D. *The Biology of Moral Systems*. Hawthorne, NY: Aldine de Gruyter, 1987.

Alper, Matthew. *The "God" Part of the Brain: A Scientific Interpretation of Human Spirituality and God*. Brooklyn, NY: Rogue Press, 2001.

Anderson, Walter Truett. *The Truth about the Truth*. New York: Jeremy P. Tarcher/Putnam, 1996.

Aristotle. *Nicomachean Ethics*. Translated by Martin Ostwald. Englewood Cliffs, NJ: Prentice-Hall, 1962.

Arnhart, Larry. *Darwinian Natural Right: The Biological Ethics of Human Nature*. Albany, NY: State University of New York Press, 1998.

Asch, Solomon. *Social Psychology*. Englewood Cliffs, NJ: Prentice-Hall, 1952.

Atkatz, David. "Quantum Cosmology for Pedestrians." *American Journal of Physics* 62 (1994): 619–27.

Atkatz, David, and Heinz Pagels. "Origin of the Universe as Quantum Tunneling Event." *Physical Review* D25 (1982): 2065–73.

Axelrod, Robert. *The Evolution of Cooperation*. New York: Basic Books, 1984.

Baggini, Julian, and Jeremy Stranghorn. *What Philosophers Think*. London: Continuum, 2003.

Ball, Philip. *The Self-Made Tapestry: Pattern Formation in Nature*. New York, Oxford: Oxford University Press, 1999.

———. *Critical Mass: How One Thing Leads to Another*. New York: Farrar, Straus and Giroux, 2004.

Barrett, Justin L. *Why Would Anyone Believe in God?* Walnut Creek, CA: AltaMira Press, 2004.

Barrow, John D., and Frank J. Tipler. *The Anthropic Cosmological Principle*. Oxford: Oxford University Press, 1986.

Begley, Sharon. "Science Finds God." *Newsweek*, July 20, 1998.

Behe, Michael J. *Darwin's Black Box: The Biochemical Challenge to Evolution*. New York: Free Press, 1996.

Benson H., J. A. Dusek, J. B. Sherwood, P. Lam, C. F. Bethea, et al. "Study of the Therapeutic Effects of Intercessory Prayer (STEP) in Cardiac Bypass Patients: A Multicenter Randomized Trial of Uncertainty and Certainty of Receiving Intercessory Prayer." *American Heart Journal* 151, no. 4 (2006): 934–42.

Bishop, Jeffrey P., and Victor J. Stenger. "Retroactive Prayer: Lots of History, Not Much Mystery, and No Science." *British Medical Journal* 329 (2004): 1444–46.

Blackmore, Susan. *Dying to Live: Near-Death Experiences.* Amherst, NY: Prometheus Books, 1993.

Blaker, Kimberly, ed. *The Fundamentals of Extremism: The Christian Right in America.* New Boston, MI: New Boston Books, 2003.

Blanke, Olaf, Stephanie Ortigue, Theodore Landis, and Margritta Seeck. "Stimulating Illusory Own-Body Perceptions." *Nature* 419 (September 19, 2002): 269–70.

Bloom, Paul. *Descartes' Baby: How the Science of Child Development Explains What Makes Us Human.* New York: Basic Books, 2004.

——. "Is God an Accident?" *Atlantic* 296, no. 5 (December 2005): 105–12.

Bohm, David, and B. J. Hiley. *The Undivided Universe: An Ontological Interpretation of Quantum Mechanics.* London: Routledge, 1993.

Boyer, Pascal. *Religion Explained: The Evolutionary Origin of Religious Thought.* New York: Basic Books, 2001.

Brauer, Matthew J., Barbara Forrest, and Steven G. Gey. "Is It Science Yet?: Intelligent Design Creationism and the Constitution." *Washington University Law Quarterly* 83, no. 1 (2005), http://law.wustl.edu/WULQ/83-1/p%201%20Brauer%20Forrest%20Gey%20book%20pages.pdf (accessed December 28, 2005).

Broom, Donald M. *The Evolution of Morality and Religion.* Cambridge: Cambridge University Press, 2003.

Brown, Warren S., Nancey Murphy, and H. Newton Malony, eds. *Whatever Happened to the Soul? Scientific and Theological Portraits of Human Nature.* Minneapolis: Fortress Press, 1998.

Bupp, Nathan. "Follow-up Study on Prayer Therapy May Help Refute False and Misleading Information about Earlier Prayer Study." *Commission for Scientific Medicine and Mental Health*, July 22, 2005, http://csmmh.org/prayer/MANTRA.release.htm (accessed December 16, 2005).

Byers, Nina. "E. Noether's Discovery of the Deep Connection between Symmetries and Conservation Laws." *Israel Mathematical Conference Proceedings* 12 (1999), http://www.physics.ucla.edu/~cwp/articles/noether.asg/noether.html (accessed July 1, 2006).

Byrd, Randolph C. "Positive Therapeutic Effects of Intercessory Prayer in a Coronary Care Unit Population." *Southern Medical Journal* 81, no. 7 (1988): 826–29.

Callahan, Tim. *Bible Prophecy: Failure or Fulfillment*. Altadena, CA: Millennium Press, 1997.

Campbell, Alexander. "Our Position to American Slavery—No. V." *Millennial Harbinger*, ser. 3, vol. 2 (1845): 193.

Carnap, Rudolf. "Testability and Meaning." *Philosophy of Science* B 3 (1936): 19–21; B 4 (1937): 1–40.

Carr, B. J., and M. J. Rees. "The Anthropic Principle and the Structure of the Physical World." *Nature* 278 (1979): 606–12.

Carrier, Richard. "The Real Ten Commandments." Internet Infidels Library (2000), http://www.infidels.org/library/modern/features/2000/carrier2.html (accessed August 14, 2005).

Carter, Brandon. "Large Number Coincidences and the Anthropic Principle in Cosmology." In *Confrontation of Cosmological Theory with Astronomical Data*, edited by M. S. Longair, 291–98. Dordrecht: Reidel, 1974. Reprinted in *Modern Cosmology and Philosophy*, edited by John Leslie, 131–39. Amherst, NY: Prometheus Books, 1998.

Cha, K. Y., D. P. Wirth, and R. A. Lobo. "Does Prayer Influence the Success of In Vitro Fertilization-Embryo Transfer? Report of a Masked, Randomized Trial." *Journal of Reproductive Medicine* 46, no. 9 (September 2001): 781–87.

Chopra, Deepak. *Quantum Healing: Exploring the Frontiers of Mind/Body Medicine*. New York: Bantam, 1989.

———. *Ageless Body, Timeless Mind: The Quantum Alternative to Growing Old*. New York: Random House, 1993.

Churchland, Patricia Smith. *Neurophilosophy: Toward a Unified Science of the Mind/Brain*. Cambridge, MA: MIT Press, 1996.

Churchland, Paul M. *The Engine of Reason, the Seat of the Soul: A Philosophical Journey into the Brain*. Cambridge, MA: MIT Press, 1996.

Cicero, Marcus Tullius. *De Natura Deorum* or *On the Nature of the Gods*. Edited and translated by H. Rackham. New York: Loeb Classical Library, 1933.

Cowen, J. L. "The Paradox of Omnipotence Revisited." *Canadian Journal of Philosophy* 3, no. 3 (March 1974): 435–45. Reprinted in *The*

Impossibility of God, edited by Michael Martin and Ricki Monnier. Amherst, NY: Prometheus Books, 2003.

Craig, William Lane. *The Kalām Cosmological Argument*. Library of Philosophy and Religion. London: Macmillan, 1979.

———. *The Cosmological Argument from Plato to Leibniz*. Library of Philosophy and Religion. London: Macmillan, 1980.

———. "The Historicity of the Empty Tomb of Jesus." *New Testament Studies* 31 (1985): 39–67, http://www.leaderu.com/offices/billcraig/docs/tomb2.html (accessed January 4, 2005).

———. *Reasonable Faith*. Wheaton, IL: Crossway, 1994.

Craig, William Lane, and Quentin Smith. *Theism, Atheism, and Big Bang Cosmology*. Oxford: Clarendon Press, 1997.

———. "The Absurdity of Life without God," http://www.hisdefense.org/audio/wc_audio.html (accessed March 9, 2004).

Dalai Lama. *The Universe in a Single Atom: The Convergence of Science and Spirituality*. New York: Random House, 2005.

Darling, David J. *Life Everywhere: The Maverick Science of Astrobiology*. New York: Basic Books, 2001.

Darwin, Charles. *The Origin of Species by Means of Natural Selection*. London: John Murray, 1859.

———. *The Correspondence of Charles Darwin* 8, 1860. Cambridge: Cambridge University Press, 1993.

Davies, Paul. *The Cosmic Blueprint*. New York: Simon and Schuster, 1988; Radnor, PA: Templeton Foundation Press, 2004.

———. "Multiverse or Design: Reflections on a Third Way." Proceedings of *Universe or Multiverse?* Stanford University, March 2003, http://aca.mq.edu.au/PaulDavies/Multiverse_StanfordUniv_March2003.pdf (accessed January 4, 2005).

Davis, Jefferson. "Inaugural Address as Provisional President of the Confederacy." Montgomery, AL, February 18, 1861. *Confederate States of America Congressional Journal* 1 (1861): 64–66, quoted in Dunbar Rowland, *Jefferson Davis's Place in History as Revealed in His Letters, Papers, and Speeches*, vol. 1, 286. Jackson, MS: Torgerson Press, 1923.

Davis, J. J. "The Design Argument, Cosmic 'Fine Tuning,' and the Anthropic Principle." *Philosophy of Religion* 22 (1987): 139–50.

Dawkins, Richard. *The Blind Watchmaker: Why the Evidence of Evolution Reveals a Universe without Design.* London: Penguin Books, 1986. Paperback edition, London: Norton, 1987.

———. *River out of Eden.* New York: HarperCollins, 1995.

———. "God's Utility Function." *Scientific American* (November 1995): 85.

———. *Climbing Mount Improbable.* New York, London: Norton, 1996.

———. *Unweaving the Rainbow: Science, Delusion and the Appetite for Wonder.* Boston, New York: Houghton Mifflin, 1998.

———. *The God Delusion.* Boston, New York: Houghton Mifflin, 2006.

de Duve, Christian. *Vital Dust.* New York: Basic Books, 1995.

Dembski, William A. *The Design Inference.* Cambridge: Cambridge University Press, 1998.

———. *Intelligent Design: The Bridge between Science and Theology.* Downers Grove, IL: InterVarsity Press, 1999.

———. *No Free Lunch: Why Specified Complexity Cannot Be Purchased without Intelligence.* Lanham, MD: Rowman & Littlefield, 2002.

Dennett, Daniel. *Consciousness Explained.* Boston: Little, Brown, 1991.

———. *Breaking the Spell: Religion as a Natural Phenomenon.* New York: Viking Penguin, 2006.

Dever, William G. *Recent Archaeological Discoveries and Biblical Research.* Seattle and London: University of Washington Press, 1990.

de Wall, Frans B. M. *Good Natured: The Origins of Right and Wrong in Humans and Other Animals.* Cambridge, MS: Harvard University Press, 1996.

Doherty, Earl. *The Jesus Puzzle: Did Christianity Begin with a Mythical Christ?* Ottawa: Canadian Humanist Publications, 1999.

Dorit, Robert. Review of *Darwin's Black Box* by Michael Behe. *American Scientist* (September/October 1997).

Dossey, Larry. *Healing Words: The Power of Prayer and the Practice of Medicine.* San Francisco: Harper, 1993.

———. Response to letter to the editor. *Southern California Physician* (December 2001): 46.

Douady, S., and Y. Couder. "Phyllotaxis as a Physical Self-Organized Growth Process," *Physical Review Letters* 68 (1992): 2098.

Drange, Theodore M. *Nonbelief and Evil: Two Arguments for the Nonexistence of God.* Amherst, NY: Prometheus Books, 1998.

———. "Incompatible-Properties Arguments—A Survey." *Philo* 1, no. 2 (1998): 49–60. In *The Impossibility of God*, edited by Michael Martin and Ricki Monnier, 185–97. Amherst, NY: Prometheus Books, 2003.

Edis, Taner. "Darwin in Mind: 'Intelligent Design' Meets Artificial Intelligence." *Skeptical Inquirer* 25, no. 2 (2001): 35–39.

Elbert, Jerome W. *Are Souls Real?* Amherst, NY: Prometheus Books, 2000.

Eller, David. *Natural Atheism*. Cranford, NJ: American Atheist Press, 2004.

Ellis, George. *Before the Beginning: Cosmology Explained*. London, New York: Boyars/Bowerdean, 1993.

Ely, Melvin Patrick. *Israel on the Appomattox: A Southern Experiment in Black Freedom from the 1790s through the Civil War*. New York: Alfred A. Knopf, 2005.

Everitt, Nicholas. *The Non-Existence of God*. London, New York: Routledge, 2004.

Fales, Evan. "Despair, Optimism, and Rebellion," http://www.infidels .org/library/modern/evan_fales/despair.html (accessed July 6, 2005).

Faraoni, V., and F. I. Cooperstock. "On the Total Energy of Open Friedmann-Robertson-Walker Universes." *Astrophysical Journal* 587 (2003): 483–86.

Ferguson, Everitt. *Background of Early Christianity*. Third edition. Grand Rapids, MI: W. B. Eerdmans, 2003.

Fernald, R. D. "Evolution of Eyes." *Current Opinions in Neurobiology* 10, no. 4 (2000): 444–50.

Finkelstein, Israel, and Neil Asher Silberman. *The Bible Unearthed: Archaeology's New Vision of Ancient Israel and the Origin of Its Sacred Texts*. New York: Free Press, 2001.

Fitelson, Brandon, Christopher Stephens, and Elliott Sober. "How Not to Detect Design—Critical Notice: William A. Dembski, The Design Inference." *Philosophy of Science* 66, no. 3 (1999): 472–88.

Flack, Jessica C., and Frans B. M. de Wall. "'Any Animal Whatever' Darwinian Building Blocks of Morality in Monkeys and Apes." *Journal of Consciousness Studies* 7, nos. 1–2 (2000): 1–29.

Flamm, Bruce L. "Faith Healing by Prayer." Review of "Does Prayer Influence the Success of In Vitro Fertilization-Embryo Transfer? Report of a Masked, Randomized Trial," by K. Y. Cha, D. P. Wirth, and R. A. Lobo, *Scientific Review of Alternative Medicine* 6, no. 1 (2002): 47–50.

———. "Faith Healing Confronts Modern Medicine." *Scientific Review of Alternative Medicine* 8, no. 1 (2004): 9–14.

———. "The Columbia 'Miracle' Study: Flawed and Fraud." *Skeptical Inquirer* 28, no. 5 (September/October 2004): 25–31.

Forrest, Barbara, and Paul R. Gross. *Creationism's Trojan Horse: The Wedge of Intelligent Design.* Oxford and New York: Oxford University Press, 2004.

Fox, Mark. *Religion, Spirituality, and the Near-Death Experience.* New York: Routledge, 2003.

Franklin, Michael, and Marian Hetherly. "How Fundamentalism Affects Society." *Humanist* 57 (September/October 1997): 25.

Freke, Timothy, and Peter Gandy. *The Jesus Mysteries: Was the "Original Jesus" a Pagan God?* New York: Harmony Books, 1999.

Fulmer, Gilbert. "A Fatal Logical Flaw in Anthropic Design Principle Arguments." *International Journal for Philosophy of Religion* 49 (2001): 101–10.

Furman, Richard. "Exposition of the View of the Baptists Relative to the Colored Population of the United States to the Governor of South Carolina 1822." Transcribed by T. Lloyd Benson from the original text in the South Carolina Baptist Historical Collection, Furman University, Greenville, South Carolina. http://alpha.furman.edu/~benson/docs/rcd-fmn1.htm (accessed December 1, 2004).

Gardner, Martin. "On Cellular Automata, Self-Reproduction, the Garden of Eden, and the Game of 'Life.'" *Scientific American* 224, no. 2 (1971): 112–17.

Gleason, Archer L. *Encyclopedia of Bible Difficulties.* Grand Rapids, MI: Zondervan, 2001.

Gleick, James. *Chaos: The Making of a New Science.* New York: Viking, 1987.

Glynn, Patrick. *God: The Evidence.* Rocklin, CA: Prima Publishing, 1997.

Gonzalez, Guillermo, and Jay W. Richards. *The Privileged Planet: How Our Place in the Cosmos Is Designed for Discovery.* Washington, DC: Regnery, 2004.

Goodstein, Laurie. "Intelligent Design Might Be Meeting Its Maker." Ideas and Trends, *New York Times*, December 4, 2005.

Gould, Stephen J. *The Panda's Thumb*. Norton: New York, 1980.

———. *Rocks of Ages: Science and Religion in the Fullness of Life*. New York: Ballantine, 1999.

Granqvist P., M. Fredrikson, P. Unge, A. Hagenfeldt, S. Valind, D. Larhammar, and M. Larsson. "Sensed Presence and Mystical Experiences Are Predicted by Suggestibility, Not by the Application of Transcranial Weak Complex Magnetic Fields." *Neuroscience Letters* 379, no. 1 (2005): 1–6.

Greene, Joshua D., Leigh E. Nystrom, Andrew D. Engell, John M. Darley, and Jonathan D. Cohen. "The Neural Bases of Cognitive Conflict and Control in Moral Judgment." *Neuron* 44 (2004): 389–400.

Gribbon, John. *Deep Simplicity: Bringing Order to Chaos and Complexity*. New York: Random House, 2004.

Guminski, Arnold. "The Kalam Cosmological Argument: The Questions of the Metaphysical Possibility of an Infinite Set of Real Entities." *Philo* 5, no. 2 (Fall/Winter 2002): 196–215.

Guth, Alan. *The Inflationary Universe*. New York: Addison-Wesley, 1997.

Guthrie, Stewart Elliott. *Faces in the Clouds: A New Theory of Religion*. New York, Oxford: Oxford University Press, 1993.

Haack, Susan. *Defending Science—within Reason*. Amherst, NY: Prometheus Books, 2003.

Hamer, Dean H. *The God Gene: How Faith Is Hardwired into Our Genes*. New York: Doubleday, 2004.

Harnik, Roni, Graham D. Kribs, and Gilad Perez. "A Universe without Weak Interactions." *Physical Review* D74 (2006): 035006.

Harris, Sam. *The End of Faith: Religion, Terror, and the Future of Reason*. New York: Norton, 2004.

———. *Letter to a Christian Nation*. New York: Alfred A. Knopf, 2006.

Harris, W. S., M. Gowda, J. W. Kolb, C. P. Strychacz, J. L. Vacek, P. G. Jones, A. Forker, J. H. O'Keefe, and B. D. McCallister. "A Randomized, Controlled Trial of the Effects of Remote, Intercessory Prayer on Outcomes in Patients Admitted to the Coronary Care Unit." *Archives of Internal Medicine* 159 (1999): 2273–78.

Hartle, J. B., and S. W. Hawking. "Wave Function of the Universe." *Physical Review* D28 (1983): 2960–75.

Haught, James A. *Holy Horrors: An Illustrated History of Religious Murder and Madness*. Amherst, NY: Prometheus Books, 1990.

Hauser, Marc, and Peter Singer. "Morality without Religion." *Free Inquiry* 26, no. 1 (December 2005/January 2006): 18–19.

Hawking, Stephen W. *A Brief History of Time: From the Big Bang to Black Holes.* New York: Bantam, 1988.

Hawking, Steven W., and Roger Penrose. "The Singularities of Gravitational Collapse and Cosmology." *Proceedings of the Royal Society of London* A, 314 (1970): 529–48.

Helms, Randel. *Gospel Fictions.* Amherst, NY: Prometheus Books, 1988.

Hoffmann, Joseph R., and Gerald A. Larue, eds. *Jesus in History and Myth.* Amherst, NY: Prometheus Books, 1986.

Hogan, Craig J. "Why the Universe Is Just So." *Reviews of Modern Physics* 72 (2000): 1149–61.

Hoyle, F., D. N. F. Dunbar, W. A. Wensel, and W. Whaling, "A State in C12 Predicted from Astrophysical Evidence." *Physical Review Letters* 92 (1953): 1095.

Huemer, Michael. "Some Failed Responses to the Problem of Evil." Talk at the University of Colorado Theology Forum, February 16, 2005, Boulder, Colorado.

Ionnidas, John P. A. "Why Most Published Research Findings Are False." *Public Library of Science, Medicine* 2, no. 8 (2005). http://medicine .plosjournals.org/perlserv/?request=get-document&doi=10.1371/ journal.pmed.0020124 (accessed December 2, 2005).

John Paul II. Address to the Academy of Sciences, October 28, 1986. *L'Osservatore Romano.* English edition. November 24, 1986.

Johnson, Phillip E. *Evolution as Dogma: The Establishment of Naturalism.* Dallas, TX: Haughton Publishing Co., 1990.

———. *Darwin on Trial.* Downers Grove, IL: InterVarsity Press, 1991.

———. *Reason in the Balance: The Case Against Naturalism in Science, Law, and Education.* Downers Grove, IL: InterVarsity Press, 1995.

———. *Defeating Darwinism by Opening Minds.* Downers Grove, IL: InterVarsity Press, 1997.

———. *The Wedge of Truth: Splitting the Foundations of Naturalism.* Downers Grove, IL: InterVarsity Press, 2001.

Johnson, Phillip E., and Howard van Till. "God and Evolution: An Exchange." *First Things* 34 (1993): 32–41.

Johnson, Steven. *Emergence: The Connected Lives of Ants, Brains, Cities, and Software.* New York: Touchstone, 2001.

Johnson, Timothy. "Praying for Pregnancy: Study Says Prayer Helps Women Get Pregnant." ABC Television. *Good Morning America,* October 4, 2001.

Kane, Gordon L., Michael J. Perry, and Anna N. Zytkow. "The Beginning of the End of the Anthropic Principle." *New Astronomy* 7 (2002): 45–53.

Katz, Leonard D., ed. *Evolutionary Origins of Morality: Cross-Disciplinary Perspectives.* Bowling Green, OH: Imprint Academic, 2000.

Kauffman, Stuart. *At Home in the Universe: The Search for the Laws of Self-Organization and Complexity.* New York and Oxford: Oxford University Press, 1995.

Kaufmann, Walter. *The Faith of a Heretic.* Paperback edition. New York: Doubleday, 1963.

Kirsch, Jonathan. *God Against the Gods: The History of the War between Monotheism and Polytheism.* New York: Viking Compass, 2004.

Kitcher, Philip J. *Abusing Science: The Case Against Creationism.* Cambridge, MA: MIT Press, 1982.

Klee, Robert. "The Revenge of Pythagoras: How a Mathematical Sharp Practice Undermines the Contemporary Design Argument in Astrophysical Cosmology." *British Journal for the Philosophy of Science* 53 (2002): 331–54.

Krauss, Lawrence. *Quintessence: The Mystery of the Missing Mass in the Universe.* New York: Basic Books, 2000.

Kristof, Nicholas D. "God and Evolution." *New York Times,* February 12, 2005, op-ed.

Krucoff, M. W., S. W. Crater, et al. "Music, Imagery, Touch, and Prayer as Adjuncts to Interventional Cardiac Care: The Monitoring and Actualization of Noetic Trainings (MANTRA) II Randomized Study." *Lancet* 366 (July 16, 2005): 211–17.

Kuhn, Thomas. *The Structure of Scientific Revolutions.* Chicago: University of Chicago Press, 1970.

Kurtz, Paul. *Forbidden Fruit: The Ethics of Humanism.* Amherst, NY: Prometheus Books, 1988.

Kushner, Harold S. *When Bad Things Happen to Good People.* New York: Avon Books, 1987.

Lakoff, George, and Mark Johnson. *Philosophy in the Flesh: The Embodied Mind and Its Challenge to Western Thought.* New York: Basic Books, 1999.

Lamont, Corliss. *The Illusion of Immortality*. Fifth edition. New York: Continuum, 1990. First published in 1935.

Larson, Edward J. and Larry Witham. "Leading Scientists Still Reject God." *Nature* 394 (1998): 313.

Leibovici, Leonard. "Alternative (Complementary) Medicine: A Cuckoo in the Nest of Empiricist Reed Warblers." *British Medical Journal* 319 (1999): 1629–31.

———. "Effects of Remote, Retroactive Intercessory Prayer on Outcomes in Patients with Bloodstream Infections: A Controlled Trial." *British Medical Journal* 323 (2001): 1450–51.

Leslie, John, ed. *Modern Cosmology and Philosophy*. Amherst, NY: Prometheus Books, 1998.

Linde Andre. "Quantum Creation of the Inflationary Universe." *Lettere Al Nuovo Cimento* 39 (1984): 401–405.

Livio, M., D. Hollowell, A. Weiss, and J. Truran. "The Anthropic Significance of the Existence of an Excited State of ^{12}C." *Nature* 340 (1989): 281–84.

Loftus, Elizabeth F. *Eyewitness Testimony*. Cambridge, MA: Harvard University Press, 1996.

Longair, M. S., ed. *Confrontation of Cosmological Theory with Astronomical Data*. Dordrecht: Reidel, 1974.

Lutz, Antonie, Lawrence L. Greischar, Nancy B. Rawlings, Mathew Ricard, and Richard J. Davidson. "Long-term Meditators Self-induce High-amplitude Gamma Synchrony during Mental Practice." *Proceedings of the National Academy of Sciences* 101, no. 46 (2004): 16369–73.

Mackie, J. J. "Evil and Omnipotence." *Mind* 64 (1955): 200–12. Reprinted in *The Impossibility of God*, edited by Michael Martin and Ricki Monnier, 61–105. Amherst, NY: Prometheus Books, 2003.

Manson, Neil A. "There Is No Adequate Definition of 'Fine-tuned for Life,'" *Inquiry* 43 (2000): 341–52.

Markale, Jean. *Montségur and the Mystery of the Cathars*. Translated by Jon Graham. Rochester, VT: Inner Traditions, 2003.

Martin, Michael, and Ricki Monnier, eds. *The Impossibility of God*. Amherst, NY: Prometheus Books, 2003.

McCabe, Joseph. *The Sources of Morality of the Gospels*. London: Watts and Co., 1914.

McDowell, Josh. *Evidence That Demands a Verdict*. San Bernardino, CA: Here's Life Publishers, 1972, 1979.

Meyer, Stephen C. "The Origin of Biological Information and the Higher Taxonomic Categories." *Proceedings of the Biological Society of Washington* 117, no. 2 (2004): 213–39.

Miller, Kenneth R. "Life's Grand Design." *Technology Review* 97, no. 2 (1994): 24–32.

———. *Finding Darwin's God: A Scientist's Search for a Common Ground between God and Evolution*. New York: HarperCollins, 1999.

Miller, Ruth, Larry S. Miller, and Mary R. Langenbrunner. "Religiosity and Child Sexual Abuse: A Risk Factor Assessment." *Journal of Child Sexual Abuse* 6, no. 4 (1997): 14–34.

Miller, Stanley L. "A Production of Amino Acids under Possible Primitive Earth Conditions." *Science* 117 (1953): 528–29.

Moll, Jorge, Ricardo de Oliveira-Souza, Ivanel E. Bramati, and Jordan Grafman. "Functional Networks in Emotive and Nonmoral Social Judgments." *NeuroImage* 16 (2002): 696–703.

Mongan, T. R. "Simple Quantum Cosmology: Vacuum Energy and Initial State." *General Relativity and Gravitation* 37 (2005): 967–70.

Monkerud, Don. "Faith No More." *In These Times* 27, no. 9 (March 10, 2003). http://inthesetimes.com/comments.php?id=105_0_2_0_C (accessed February 19, 2005).

Mooney, Chris. *The Republican War on Science*. New York: Perseus Books Group, 2005.

Moreland, J. P., and Kai Nielsen. *Does God Exist? The Debate between Theists & Atheists*. Amherst, NY: Prometheus Books, 1993.

Morriston, Wes. "Creation *Ex Nihilo* and the Big Bang." *Philo* 5, no. 1 (2002): 23–33.

Muller, H. J. "Reversibility in Evolution Considered from the Standpoint of Genetics." *Biological Reviews* 14 (1939): 261–80.

Murphy, Dean E., and Neela Banjeree. "Catholics in U.S. Keep Faith, but Live With Contradictions." *New York Times*, April 11, 2005.

Nagel, Thomas. *Mortal Questions*. Cambridge: Cambridge University Press, 1979.

Nakamura, Takashi, H. Uehara, and T. Chiba. "The Minimum Mass of the First Stars and the Anthropic Principle." *Progress of Theoretical Physics* 97 (1997): 169–71.

National Academy of Sciences. *Teaching About Evolution and the Nature of Science*. Washington, DC: National Academy of Sciences, 1998, http://www.nap.edu/catalog/5787.html (accessed March 5, 2006).

Nelson-Pallmeyer, Jack. *Is Religion Killing Us? Violence in the Bible and the Quran*. Harrisburg, PA: Trinity Press International, 2003.

Newberg, Andrew, and Eugene d'Aquili. *Why God Won't Go Away*. New York: Ballantine Books, 2001.

Nguyen, Tommy. "Smithsonian Distances Itself from Controversial Film." *Washington Post*, June 2, 2005.

Nickell, Joe. *Inquest on the Shroud of Turin*. Amherst, NY: Prometheus Books, 1987.

———. "Bone (Box) of Contention: The James Ossuary." *Skeptical Inquirer* 27, no. 2 (March/April 2003): 19–22.

Nielsen, Kai. *Ethics without God*. Revised edition. Amherst, NY: Prometheus Books, 1990.

Noether, E. "Invarianten beliebiger Differentialausdrücke." *Nachr. d. König. Gesellsch. d. Wiss. zu Göttingen, Math-phys.* Klasse (1918): 37–44. See Nina Byers, "E. Noether's Discovery of the Deep Connection between Symmetries and Conservation Laws," *Israel Mathematical Conference Proceedings* 12 (1999), http://www.physics.ucla.edu/~cwp/articles/noether.asg/noether.html (accessed July 1, 2006), for English translation.

Noonan, John T., Jr. *A Church That Can and Cannot Change: The Development of Catholic Moral Teaching*. Notre Dame, IN: University of Notre Dame Press, 2005.

Numbers, Ronald. *The Creationists: The Evolution of Scientific Creationism*. New York: Alfred A. Knopf, 1992.

Oden, J. Tinsley. Acceptance remarks, 1993 John von Neumann Award Winner. *United States Association of Computational Mechanics Bulletin* 6, no. 3 (September 1993), http://www.usacm.org/Oden's_acceptance_remarks.htm (accessed February 22, 2005).

Olshansky, Brian, and Larry Dossey. "Retroactive Prayer: a Preposterous Hypothesis?" *British Medical Journal* 327 (2003): 1460–63.

Olshansky, S. Jay, Bruce Carnes, and Robert N. Butler. "If Humans Were Built to Last." *Scientific American* (March 2001).

Oppy, Graham. "Arguing *About* the *Kalam* Cosmological Argument." *Philo* 5, no. 1 (Spring/Summer 2002): 34–61.

Orr, H. Allen. "Darwin v. Intelligent Design (Again): The Latest Attack on Evolution Is Cleverly Argued, Biologically Informed—And Wrong." *Boston Review* (1998).

Overman, Dean L. *A Case Against Accident and Self-Organization*. New York, Oxford: Rowman & Littlefield, 1997.

Overton, William R. *McLean v. Arkansas*, United States District Court Opinion, 1982.

Paley, William. *Natural Theology or Evidences of the Existence and Attributes of the Deity Collected from the Appearance of Nature*. London: Halliwell, 1802.

Parsons, Keith. *God and the Burden of Proof: Platinga, Swinburne, and the Analytical Defense of Theism*. Amherst, NY: Prometheus Books, 1989.

———. "Is There a Case for Christian Theism?" in *Does God Exist? The Debate between Theists & Atheists*, by J. P. Moreland and Kai Nielsen. Amherst, NY: Prometheus Books, 1993.

Paul, Gregory S. "The Great Scandal: Christianity's Role in the Rise of the Nazis." *Free Inquiry* 23, no. 4 (October/November 2003): 20–29; 24, no. 1 (December 2003/January 2004): 28–34.

Pennock, Robert T. *Tower of Babel: The Evidence Against the New Creationism*. Cambridge, MA: MIT Press, 1999.

Perakh, Mark. "Not a Very Big Bang about Genesis" (December 2001). Online at *Talk Reason*, http://www.talkreason.org/articles/schroeder .cfm (accessed December 15, 2004).

———. *Unintelligent Design*. Amherst, NY: Prometheus Books, 2003.

Perlmutter, S., G. Aldering, G. Goldhaber, R. A. Knop, P. Nugent, P. G. Castro, S. Deustua, et al. "Measurements of Omega and Lambda from 42 High-Redshift Supernovae." *Astrophysical Journal* 517 (1999): 565–86.

Persinger, Michael A. "Paranormal and Religious Beliefs May Be Mediated Differently by Subcortical and Cortical Phenomenological Process of the Temporal (Limbic) Lobes." *Perceptual and Motor Skills* 76 (1993): 247–51.

Petre, Jonathan. "Power of Prayer Found Wanting in Hospital Trial." *News Telegraph*, October 15, 2003. http://news.telegraph.co.uk/ news/main.jhtml?xml=/news/2003/10/15/npray15.xml (accessed December 6, 2004).

Pius XII. *Humani Generis*. August 12, 1950.

Pius XII. "The Proofs for the Existence of God in the Light of Modern Natural Science." Address of Pope Pius XII to the Pontifical Academy of Sciences, November 22, 1951. Reprinted as "Modern Science and the Existence of God." *Catholic Mind* 49 (1972): 182–92.

Popper, Karl. *The Logic of Scientific Discovery*. English edition. London: Hutchinson; New York: Basic Books, 1959. Originally published in German. Vienna: Springer Verlag, 1934.

———. "Natural Selection and the Emergence of Mind." *Dialectica* 32 (1978): 339–55.

———. "Metaphysics and Criticizability." In *Popper Selections*, edited by David Miller. Princeton, NJ: Princeton University Press, 1985. Originally published in 1958.

Poundstone, William. *The Rescursive Universe*. New York: Morrow, 1985.

Press, W. H., and A. P. Lightman. "Dependence of Macrophysical Phenomena on the Values of the Fundamental Constants." *Philosophical Transactions of the Royal Society of London* A 310 (1983): 323–36.

Pullman, Bernard. *The Atom in the History of Human Thought*. Oxford: Oxford University Press, 1998.

Rachels, James. "God and Moral Autonomy." In *Can Ethics Provide Answers? And Other Essays in Moral Philosophy*. New York: Rowman & Littlefield, 1997. Reprinted in *The Impossibility of God*, edited by Michael Martin and Ricki Monnier, 45–58. Amherst, NY: Prometheus Books, 2003.

Rachels, James, and David Roche. "A Bit Confused: Creationism and Information Theory." *Skeptical Inquirer* 25, no. 2 (2001): 40–42.

Radin, Dean. *The Conscious Universe: The Scientific Truth of Psychic Phenomena*. New York: HarperEdge, 1997.

Ramachandran, V. S. "God and the Temporal Lobes of the Brain." Talk at the conference Human Selves and Transcendental Experiences: A Dialogue of Science and Religion, University of California, San Diego, January 31, 1998.

Reiss, A., A. V. Filippenko, P. Challis, A. Clocchiattia, A. Diercks, P. M. Garnavich, R. L. Gilliland, et al. "Observational Evidence from Supernovae for an Accelerating Universe and a Cosmological Constant." *Astronomical Journal* 116 (1998): 1009–38.

Ross, Hugh. *The Creator and the Cosmos: How the Greatest Scientific Dis-*

coveries of the Century Reveal God. Revised edition. Colorado Springs: Navpress, 1995. First published in 1993.

———. "Astronomical Evidences for a Personal, Transcendent God." In *The Creation Hypothesis*, edited by J. P. Moreland, 41–72. Downers Grove, IL: InterVarsity Press, 1994.

Rowland, Dunbar. *Jefferson Davis's Place in History as Revealed in His Letters, Papers, and Speeches.* Jackson, MS: Torgerson Press, 1923.

Rundle, Bede. *Why There Is Something Rather Than Nothing.* Oxford: Clarendon Press, 2004.

Ruse, Michael, ed. *But Is It Science? The Philosophical Questions in the Creation/Evolution Controversy.* Amherst, NY: Prometheus Books, 1996.

Sagan, Carl. *Pale Blue Dot.* London: Headliner, 1995.

Salpeter, E. E. "Accretion of Interstellar Matter by Massive Objects." *Astrophysical Journal* 140 (1964): 796–800.

Scalia, Antonin. "God's Justice and Ours." *First Things* 123 (May 2002): 17–21. http://www.firstthings.com/ftissues/ft0205/articles/scalia .html (accessed March 15, 2005).

Schellenberg, John L. *Divine Hiddenness and Human Reason.* Ithaca, NY: Cornell University Press, 1993.

Schick, Theodore, Jr. "Is Morality a Matter of Taste? Why Professional Ethicists Think That Morality Is *Not* Purely Subjective." *Free Inquiry* 18, no. 4 (1998): 32–34.

Schroeder, Gerald L. *Genesis and the Big Bang. The Discovery of the Harmony between Modern Science and the Bible.* New York: Bantam Books, 1992.

———. *The Science of God: The Convergence of Scientific and Biblical Wisdom.* New York: Broadway Books, 1998.

———. *The Hidden Face of God: How Science Reveals the Ultimate Truth.* New York: Free Press, 2001.

Schuster, Angela M. H. "Not Phillip II of Macedon." *Archaeology* (April 20, 2000). http://www.archaeology.org/online/features/macedon/ (accessed December 26, 2004).

Shallit, Jeffery. Review of *No Free Lunch* by William Dembski. *Biosystems* 66, nos. 1–2 (2002): 93–99.

Shanks, Niall, and Karl H. Joplin. "Redundant Complexity: A Critical Analysis of Intelligent Design in Biochemistry." *Philosophy of Science* 66 (1999): 268–98.

Shannon, C. E. "A Mathematical Theory of Communication." *Bell System Technical Journal* 27 (July 1948): 379–423; (October 1948): 623–25.

Shannon, Claude, and Warren Weaver. *The Mathematical Theory of Communication*. Urbana: University of Illinois Press, 1949.

Shermer, Michael. *In Darwin's Shadow: The Life and Science of Alfred Russel Wallace*. Oxford, New York: Oxford University Press, 2002.

———. *The Science of Good & Evil: Why People Cheat, Gossip, Care, Share, and Follow the Golden Rule*. New York: Times Books, 2004.

Sider, Ronald J. "The Scandal of the Evangelical Conscience." *Christianity Today* 11, no. 1 (January/February 2005): 8. http://www.christianitytoday.com/bc/2005/001/3.8.html (accessed March 22, 2005).

Singer, Peter. *How Are We to Live? Ethics in an Age of Self-Interest*. Amherst, NY: Prometheus Books, 1995.

———. *The President of God and Evil: The Ethics of George W. Bush*. New York: Dutton, 2004.

Smith, George. *Atheism: The Case Against God*. Amherst, NY: Prometheus Books, 1989.

Stefanatos, Joanne. "Introduction to Bioenergetic Medicine." In *Complementary and Alternative Veterinary Medicine: Principles and Practice*, edited by Allen M. Schoen and Susan G. Wynn, chap. 16. St. Louis: Mosby-Year Book, 1998.

Stenger, Victor J. *Not by Design: The Origin of the Universe*. Amherst, NY: Prometheus Books, 1988.

———. *Physics and Psychics: The Search for a World beyond the Senses*. Amherst, NY: Prometheus Books, 1990.

———. *The Unconscious Quantum: Metaphysics in Modern Physics and Cosmology*. Amherst, NY: Prometheus Books, 1995.

———. "Bioenergetic Fields." *Scientific Review of Alternative Medicine* 3, no. 1 (Spring/Summer 1999).

———. *Timeless Reality: Symmetry, Simplicity, and Multiple Universes*. Amherst, NY: Prometheus Books, 2000.

———. "Natural Explanations for the Anthropic Coincidences." *Philo* 3, no. 2 (2001): 50–67.

———. *Has Science Found God? The Latest Results in the Search for Purpose in the Universe*. Amherst, NY: Prometheus Books, 2003.

———. "Fitting the Bible to the Data." *Skeptical Inquirer* 23, no. 4 (1999): 67. Online at *Secular Web*, http://www.infidels.org/library/modern/vic_stenger/schrev.html (accessed December 13, 2004).

———. *The Comprehensible Cosmos: Where Do the Laws of Physics Come From?* Amherst, NY: Prometheus Books, 2006.

Sterne, Jonathan A., and George Davey Smith. "Sifting the Evidence— What's Wrong with Significance Tests?" *British Medical Journal* 322 (2001): 226–31.

Stokes, Douglas M. "The Shrinking Filedrawer: On the Validity of Statistical Meta-Analysis in Parapsychology." *Skeptical Inquirer* 25, no. 3 (2001): 22–25.

Swinburne, Richard. *The Existence of God.* Oxford: Clarendon Press, 1979.

———. "Argument from the Fine-Tuning of the Universe." In *Modern Cosmology and Philosophy*, edited by John Leslie, 160–79. Amherst, NY: Prometheus Books, 1998.

Tart, Charles T. "A Psychophysiological Study of Out-of-the-Body Experiences in a Selected Subject." *Journal of the American Society for Psychical Research* 62 (1968): 3–27.

Tegmark, Max. "Does the Universe in Fact Contain Almost No Information?" *Foundations of Physics Letters* 9, no. 1 (1996): 25–42.

Thomson, Keith. *Before Darwin: Reconciling God and Nature.* New Haven and London: Yale University Press, 2005.

Thorne, Kip S. *Black Holes & Time Warps: Einstein's Outrageous Legacy.* New York: Norton, 1994.

Totten, R. "The Intelligent Design of the Cosmos: A Mathematical Proof" (2000). http://www.geocities.com/worldview_3/mathprfcosmos.html (accessed February 6, 2005).

Tryon, E. P. "Is the Universe a Quantum Fluctuation?" *Nature* 246 (1973): 396–97.

Ussery, David. "Darwin's Transparent Box: The Biochemical Evidence for Evolution." In *Why Intelligent Design Fails: A Scientific Critique of the New Creationism*, edited by Matt Young and Taner Edis, chap. 4. New Brunswick, NJ, and London: Rutgers University Press, 2004.

Vilenkin, Alexander. "Birth of Inflationary Universes." *Physical Review* D27 (1983): 2848–55.

———. "Quantum Creation of Universes." *Physical Review* D30 (1984): 509.

Walton, Douglas. "Can an Ancient Argument of Carneades on Cardinal Virtues and Divine Attributes Be Used to Disprove the Existence of God?" *Philo* 2, no. 2 (1999): 5–13. Reprinted in *The Impossibility of God*, edited by Michael Martin and Ricki Monnier, 35–44. Amherst, NY: Prometheus Books, 2003.

Ward, Peter D., and Donald Brownlee. *Rare Earth: Why Complex Life Is Uncommon in the Universe*. New York: Copernicus, 2000.

Weinberg, Steven. "The Cosmological Constant Problem." *Reviews of Modern Physics* 61 (1989): 1–23.

———. "The Revolution That Didn't Happen." *New York Review of Books*, October 8, 1998.

———. "A Designer Universe?" *New York Review of Books*, October 21, 1999. Reprinted in the *Skeptical Inquirer* (September/October 2001): 64–68.

Wells, G. A. *The Historical Evidence for Jesus*. Amherst, NY: Prometheus Books, 1988.

Whitcomb, John C., Jr., and Henry M. Morris. *The Genesis Flood: The Biblical Record and Its Scientific Implications*. Philadelphia: Presbyterian and Reformed Publishing Co., 1961.

Wielenberg, Erik J. *Value and Virtue in a Godless Universe*. Cambridge, New York: Cambridge University Press, 2005.

Wilczek, Frank. "The Cosmic Asymmetry between Matter and Anti-matter." *Scientific American* 243, no. 6 (1980): 82–90.

Will, Clifford M. *Was Einstein Right? Putting General Relativity to the Test*. New York: Basic Books, 1986.

Wolfram, Stephen. *A New Kind of Science*. Champagne, IL: Wolfram Media, 2002.

Wright, Robert. *The Moral Animal: Why We Are the Way We Are: The New Science of Evolutionary Psychology*. New York: Vintage Books, 1994.

Xin Yan, F. Lu, H. Jiang, X. Wu, W. Cao, Z. Xia, H. Shen, et al. "Certain Physical Manifestation and Effects of External Qi of Yan Xin Life Science Technology." *Journal of Scientific Exploration* 16, no. 3 (2002): 381–411.

Young, Matt, and Taner Edis, eds. *Why Intelligent Design Fails: A Scientific Critique of the New Creationism*. New Brunswick, NJ, and London: Rutgers University Press, 2004.

Zimmer, Carl. *Soul Made Flesh: The Discovery of the Brain—and How It Changed the World.* New York: Free Press, 2004.

Zusne, Leonard, and Warren H. Jones. *Anomalistic Psychology: A Study of Extraordinary Phenomena of Behavior and Experience.* Hillsdale, NJ: Lawrence Eribaum Associates, 1982.

INDEX

ABOUT THE AUTHOR

Victor Stenger grew up in a Catholic working-class neigh-
borhood in Bayonne, New Jersey. His father was a
Lithuanian immigrant, his mother the daughter of Hungarian
immigrants. He attended public schools and received a bachelor
of science degree in electrical engineering from Newark College
of Engineering (now New Jersey Institute of Technology) in 1956.
While at NCE, he was editor of the student newspaper and
received several journalism awards.

Moving to Los Angeles on a Hughes Aircraft Company fellow-
ship, Stenger received a master of science degree in physics from
UCLA in 1959 and a PhD in physics in 1963. He then took a
position on the faculty of the University of Hawaii, retiring to
Colorado in 2000. His current position is emeritus professor of
physics and astronomy at the University of Hawaii and adjunct

professor of philosophy at the University of Colorado. Stenger is a fellow of CSICOP and a research fellow of the Center for Inquiry. Stenger has also held visiting positions on the faculties of the University of Heidelberg in Germany, Oxford in England (twice), and has been a visiting researcher at Rutherford Laboratory in England, the National Nuclear Physics Laboratory in Frascati, Italy, and the University of Florence in Italy.

His research career spanned the period of great progress in elementary particle physics that ultimately led to the current *standard model*. He participated in experiments that helped establish the properties of strange particles, quarks, gluons, and neutrinos. He also helped pioneer the emerging fields of very high-energy gamma-ray and neutrino astronomy. In his last project before retiring, Stenger collaborated on the underground experiment in Japan that showed for the first time that the neutrino has mass.

Victor Stenger has had a parallel career as an author of critically well-received popular-level books that interface between physics and cosmology and philosophy, religion, and pseudoscience. These include: *Not By Design: The Origin of the Universe* (1988); *Physics and Psychics: The Search for a World beyond the Senses* (1990); *The Unconscious Quantum: Metaphysics in Modern Physics and Cosmology* (1995); *Timeless Reality: Symmetry, Simplicity, and Multiple Universes* (2000); *Has Science Found God? The Latest Results in the Search for Purpose in the Universe* (2003); and *The Comprehensible Cosmos: Where Do the Laws of Physics Come From?* (2006).

Stenger and his wife Phylliss have been happily married since 1962 and have two children and four grandchildren. They attribute their long lives to the response of evolution to the human need for babysitters, a task they joyfully perform. Phylliss and Vic are avid doubles tennis players, golfers, generally enjoy the outdoor life in Colorado, and travel the world as often as they can.

Stenger maintains a popular Web site (a thousand hits per month), where much of his writing can be found, at http://www.colorado.edu/philosophy/vstenger/.